**Freeman Laboratory Experiments Now Available through
Freeman's Custom Publishing Service**

Dear Instructor:
Now there is a more convenient way to have the lab manual you want, drawn
from Freeman's acclaimed archive of biology experiments. Visit Freeman's
Custom Publishing Service at http://custompub.freeman.com and create the
manual that is just right for your course. (Note: This service is now the
exclusive way to order selected individual experiments, which were formerly
available as laboratory separates. Freeman will continue to offer its lab
manuals, including the *Anatomy and Dissection* series as separate bound
volumes.)

It's Fast
Building and ordering your custom book takes just minutes. You determine
the content, organization, and cover copy. We confirm price and ISBN with
you, and the books are then printed and bound and shipped to your bookstore
4-6 weeks after the order is placed, depending on the option you choose.

It's Flexible
You can create the custom book you want--even incorporating your own
material or blank pages. The system automatically builds the book with
correct table of contents and running headers and footers. Each book you
create is stored for future printings or modifications.

It's Freeman
You will be choosing material from Freeman's rich database of content. The
manuals are printed digitally to ensure first-generation output quality. No
blurry images or illegible type!

To view our database of experiments, and for additional information, please
visit http://custompub.whfreeman.com or contact your local WH Freeman
sales representative.

FIFTH EDITION

ANATOMY & DISSECTION of the FETAL PIG

Warren F. Walker, Jr.
EMERITUS PROFESSOR, OBERLIN COLLEGE

Dominique G. Homberger
PROFESSOR, LOUISIANA STATE UNIVERSITY, BATON ROUGE

ILLUSTRATED BY

Fine Line Illustrations

W. H. FREEMAN AND COMPANY
New York

Acquisitions Editor: Patrick Shriner

Associate Editor: Debra Siegel

Assistant Editor: Melina LaMorticella

Project Editor: Sarah Kimnach, ESNE

Production Coordinator: Maura Studley

Administrative Assistant: Ceserina Pugliese

Cover and Text Design: John Svatek, ESNE

Manufacturing: Olympian Graphics

ISBN-13: 978-0-7167-2637-1
ISBN-10: 0-7167-2637-8

Printed in the United States of America

Eighth printing

T A B L E   O F

CONTENTS

O NLY A FEW REMARKS need to be made about anatomical exercises. They relate to the specimens used and to their care, to useful supplementary material, and to parts of the exercises in which students might need special help.

The fetal pig is a relatively small animal, with accordingly small organs and structures. Some structures, such as certain salivary glands, some muscles, nerves of the autonomic nervous system, smaller blood vessels, and the reproductive organs, are especially difficult to observe with the naked eye of even young students. A low-cost alternative to making dissecting microscopes available to students is magnifying lamps (3×). Various models are available at office supply and other stores for a relatively modest cost. Several of these lamps for demonstration specimens and for students to use for the smallest structures can make a big difference in the student's comprehension of the described materials.

Exercise 1
External Anatomy, Skin, and Skeleton

Preserved specimens of fetal pigs can be obtained from any biological supply company. Large specimens, ranging in length from 10 to 14 inches, should be ordered. They should be doubly injected if both the arteries and veins are to be studied in Exercise 4. Some teachers have students dissect only the arteries, and the veins are demonstrated in specially prepared specimens. Specimens are usually fixed in formalin (an aqueous solution of formaldehyde gas), because it penetrates tissues rapidly and chemically links proteins, thereby protecting them from enzymatic and bacterial decay. After fixation, specimens used to be stored in formalin, but now they usually are transferred to a holding solution based on isopropanol, ethylene glycol, or a 1–2% phenoxyethanol solution. These solutions prevent the growth of mold and associated tissue deterioration. Isopropanol, however, is very flammable, so sparks and open flames should be avoided in its presence. As specimens are dissected, they should be sprayed or doused with a wetting solu-

tion, such as a 1% phenoxyethanol solution. Phenoxyethanol is a nontoxic preserving agent with a rather pleasant scent, and it softens and rehydrates the tissues of the specimen. To prepare such a solution, mix one part of phenoxyethanol (2-phenoxyethanol practical grade, from Eastman Chemical Co.) with 99 parts of very hot tap water. Because phenoxyethanol is slightly lipophilic, you may wish to use a magnetic stirrer for about 30 minutes to ensure a homogeneous solution.

If the specimens are held in formalin, remove as much unbound formaldehyde gas as you can by thoroughly rinsing the specimens in running water. Escaping unbound formaldehyde gas irritates the skin and nose and is mildly carcinogenic, although not in the concentrations usually encountered in the laboratory. Always be cautious in handling preserved material, particularly if you have sensitive skin and respiratory problems. Drinks and foods certainly do not belong in a laboratory.

Between laboratory sessions, wrap the specimens in cloths soaked with the wetting solution, and store them in tightly closed plastic bags or boxes. If the specimens become hard and dry, add one part of any laundry fabric softener to 99 parts of the 1% phenoxyethanol solution and immerse the specimens in this. Additional information of this and other techniques can be found in the following references, which are cited in full in the References: Blaney and Johnson (1989); Frolich, Anderson, Knutsen, and Flood (1984); Rosenberg (1992a, b), and Wineski and English (1989). If specimens are held in solutions provided by the supplier, follow the supplier's instructions with respect to their care.

B. Integument

Descriptions of microscopic anatomy throughout these exercises are based on histological slides of mammalian tissue, usually from a cat or a rat. The differences between the microscopic anatomy of different mammalian species is negligible at the level at which the organs are studied. Individual histological slides may be provided to the students or a series of demonstrations may be set up. If students are given individual slides, some structures may not show in all slides; therefore, certain structures should be shown in demonstrations. The most difficult structures for students to find in skin are sweat glands, capillaries, and arrector pili muscles.

C. Skeleton

Not all institutions will have pig skeletons, but a mounted skeleton of a cat, or some other quadruped, and one of a human being should be available for comparison. General features of the skull are easier to study on separate skulls than on mounted skeletons. Vertebral structure should be studied on isolated thoracic or lumbar vertebrae of some convenient mammal.

D. Skeletal Materials

A longitudinal section of a long bone is needed if students are to see spongy and compact bone. Cartilage and bone structure may be studied on demonstration histological slides or on slides provided to each student. Slides of dried bone, thinly ground, show bone structure better than do histological sections of fixed bone.

Exercise 2 Muscles
D. to J. Muscle Groups

Muscles are described by regions of the body so that instructors can select certain groups if time does not permit a complete dissection. It is necessary to dissect the shoulder muscle before those on the cranial part of the trunk can be studied, but other groups can be studied independently.

The muscles of a fetal pig are not as well developed as in young and adult specimens. Fetal muscles are small, thin, usually very pale, and easily torn. In contrast, the connective tissue that surrounds and connects the individual muscles is well developed and often tougher and stronger than the fetal muscle tissue. Nerves leading to the muscles often form extremely strong threads that cannot be torn and must be bisected with a pair of scissors. These conditions conspire to make the dissection of fetal muscles a challenging, although rewarding exercise. The dissection is facilitated by using watchmaker's forceps as an additional tool and by performing the dissection under a magnifying lamp (3×) or a dissecting microscope. If possible, muscles should be studied in a late fetus close to term.

K. Muscle Tissue

If histological slides of muscle tissue are not available, the tissues can be seen in slides of organs: the tongue for skeletal muscle, the heart for cardiac muscle, and the intestine for smooth muscle. Muscle fiber shape can best be seen on demonstrations in which some of the individual muscle fibers are pulled apart slightly. A high-power, oil-emersion demonstration must be used to show the A and I bands clearly.

Exercise 3 Digestive and Respiratory Systems
A.1. Salivary Glands

Because salivary glands are very difficult to dissect, an instructor may wish to have students study them on demonstration dissections. Even if students do their own dissection, a demonstration will help students find the glands.

A.2. Mouth, Pharynx, Larynx, and Neck

Demonstrating to groups of students the procedure for opening the mouth and pharynx will save a lot of time and reduce the likelihood of the destruction of some structures in this region.

Certain features of the head and neck can be seen particularly well in sagittal sections, and several such sections should be demonstrated.

Vocal cords are poorly developed in a fetal pig but can be seen very clearly in a cow's larynx, which can usually be obtained from a local slaughterhouse. A "moo" can be produced by blowing through the trachea of a fresh preparation and appropriately manipulating the larynx.

B.1. Opening the Body Cavity

Again, much time can be saved by demonstrating to groups of students the procedure for opening the body cavity.

C. Respiratory and Digestive Organs of the Thorax

Look for features of a lung at this time by dissecting a lobe and examining a histological slide. The division of the trachea into bronchi should not be looked for until the heart is removed in Exercise 4.

D.2. Small Intestine and Digestive Glands

Have a demonstration available to help students find the pancreatic duct.

E. Microscopic Structure of the Small Intestine

The small intestine of the cat is described, as it is the type of histological slide most readily available; the description applies to the intestine of most mammals.

Exercise 4 Circulatory System
C. Veins and Arteries of the Neck, Head, and Thorax

The confluence of veins near the base of the neck to form the brachiocephalic veins and cranial vena cava is often difficult for students to find because the degree to which the veins are filled with the injection mass varies among specimens and because some of the veins may have been destroyed in earlier dissections. A demonstration dissection of these veins can be very helpful.

D.1. Blood Vessels of the Digestive Organs

The hepatic portal system is not intentionally injected, but in some specimens it may be injected with the blue mass introduced into the umbilical vein. Connections between branches of the umbilical vein and the hepatic portal system within the liver often carry the injection mass into the hepatic portal system. In any case, the hepatic portal veins are easily seen if the pancreas is carefully removed. A demonstration dissection would be helpful.

E. Root of the Lung and Structure of the Heart

The heart of the fetal pig is large enough to show most parts of the mammalian heart, but it is helpful to have a demonstration of a larger sheep heart so that students can see the valves clearly.

If you can obtain a fresh sheep or beef pluck (heart and lungs removed as a unit) from a local slaughter house, you can simulate with water the flow of blood through the heart, and the action of the heart valves in preventing backflow. An educational institution is usually given permission to obtain a fresh pluck in which the heart has not been cut open. Slaughter houses are normally required to slice open the heart to drain blood from it.

Dissect away the pericardial wall and other tissues to expose the major vessels entering and leaving the heart. Look for and save the ligamentum arteriosum connecting the pulmonary trunk and aortic arch. Cut open the right atrium and left ventricle, and remove any clotted blood. Also nick one pulmonary artery and one pulmonary vein.

Dissect very carefully on the left and right side of the base of the aorta to find the origin of the coronary arteries. Because the left ventricle has been cut open, the coronary arteries will leak when water is inserted into the aorta. Tie or clamp off the right coronary artery, and slip a piece of cloth beneath the left coronary artery so it can be tied off when necessary. Insert a hose line from a water tap into the brachiocephalic artery, tie it securely in place, and tie off the distal end of the aorta. A second hose line, which can be taken off the first hose line by a Y-tube, should be available for inserting into various structures. A clamp will be needed to direct water into one or other hoses.

To simulate the flow of oxygen-depleted blood into the lungs, insert the free hose through the right atrium into the right ventricle. Water (oxygen-depleted blood) will emerge through the cut pulmonary artery. Clamp off the pulmonary trunk with your hand to increase the water pressure in the right ventricle; the closed right atrioventricular valve can be seen by looking into the right atrium.

To simulate the return of oxygen-rich blood from the lungs, insert the free hose into the cut pulmonary vein and observe water entering the cut left ventricle through the left atrioventricular valve.

The action of the aortic valve can be demonstrated by running water into the hose that has been tied into the brachiocephalic artery. Clamp off the left coronary artery when you do this. The aorta will swell as pressure builds up, and the closed aortic valve can be seen by looking into the left ventricle at the point of origin of the aorta. While there is still water pressure in the aorta, release the left coronary artery and water will spurt through its branches, which were cut in opening the left ventricle. This demonstrates the separate coronary circulation.

G. Blood

If students study histological slides or smears of mammalian blood, it is unlikely that they will find all the types of leukocytes. Eosinophils and basophils are particularly hard to find and should be demonstrated, preferably under an oil-emersion objective.

Exercise 5 Urogenital System

A. Excretory System

Triply injected sheep or pig kidneys (arteries, veins, and ureter injected), obtainable from biological supply houses, show kidney structure and vascular supply very well. The distribution of renal corpuscles in the cortex can be seen with low magnification.

C. The Fetal Pig and Its Extraembyonic Membranes

This section requires a demonstration of the uterus of a pregnant sow, which can be obtained from biological supply houses. Many parts of the female genital tract will also be very clear in such a preparation, including ovarian follicles, corpora lutea, ostium tubae, uterine tube, and the parts of the uterus.

Teacher's Guide

D. Gonads

Use individual or demonstration histological slides of a mammalian testis and ovary.

Exercise 6 Nervous Coordination: Sense Organs

A. Eye

Most of the major features of the mammalian eye can be seen in the eye of the fetal pig if dissecting microscopes are used. You may prefer to use larger sheep or cow eyes. In any case, dissections of some larger eyes should be demonstrated.

B. Ear

Because it is encased by cartilage and thin bone, the middle ear of a fetal pig can be exposed exceptionally well, but demonstrations of the inner ear will be needed. Certain supply houses market a dissection of a human temporal bone in which the chambers that contain the various parts of the inner ear have been carefully exposed.

C. Nose

Because much of the skull is cartilaginous, the nose is easy to dissect in a fetal pig. However, unless students are very careful, the brain may be injured in making this dissection. You may wish to have demonstration dissections, or return to the nose after the brain has been studied. The vomeronasal organ must be looked for before the nasal septum is destroyed.

Exercise 7 Nervous Coordination: Nervous System
A.2. Microscopic Structure of the Spinal Cord and the Spinal Nerves

Histological slides of the spinal cord and spinal nerves of a small terrestrial vertebrate, such as a frog, are particularly useful. The basic structure of the spinal cord and spinal nerves is similar in all terrestrial vertebrates, and these structures are small enough in a frog that all parts can often be seen in the same microscopic field under low magnification. Mammalian slides may be used, but you may need to use different histological slides to show the spinal cord and spinal nerves.

B. Brain and Cranial Nerves

Because the brain has not been hardened in a fetal pig, specially preserved sagittal sections of sheep brains should be used if at all possible. Specimens should be checked to be sure that they are cut in the sagittal plane. A larger half can easily be dissected down to the sagittal plane. If handled carefully, the specimens can be used by many laboratory sections, and for several years. Useful demonstrations include one showing the cranial nerves, because many of these nerves are delicate and easily destroyed, another one of a brain with the dura mater intact, and another one of the brain in which the vessels in the pia mater have been injected.

If a lateral dissection is made of the pig brain (Fig. 7-4), the trigeminal, facial, vagus, and hypoglossal nerves, in particular, will be seen. A demonstration dissection would help the students identify them.

AUTHORS' Preface

D R. DOMINIQUE G. HOMBERGER of Louisiana State University, Baton Rouge, has joined the senior author in preparing the fifth edition of *Anatomy and Dissection of the Fetal Pig.* The manual is enriched by the perspective of two of us, and the participation of an active teacher helps us be aware of the needs of current students. We have prepared this edition, like its predecessors, to give introductory biology and zoology students a thorough understanding of mammalian structure and function. This compact set of laboratory exercises is designed to allow instructors maximum flexibility in course organization. It emphasizes gross anatomy but includes directions for studying the microscopic anatomy of selected organs. Some knowledge of microscopic anatomy enhances the student's understanding of gross structure and function.

A near-term fetal pig is a nice animal to study. It is readily available, of convenient size, easy to dissect, and quite representative of mammals in general. We have indicated where there are significant differences between the pig and human structure. Most of the fetal pig's organs are in the adult form, but in some cases interesting transitional stages between the fetus and adult can be observed. Instructions are included for studying the sheep eye and brain, because some instructors find them easier to examine than the comparable fetal pig organs.

Changes in the Fifth Edition

We have been guided in preparing the fifth edition by our past experience and by suggestions sent to the publisher by users of the fourth edition. Especially helpful were responses we received to certain questions the publisher asked from Dr. Roy E. Dawson, University of Colorado at Denver; Dr. Carol Feit, Yeshiva University, and an anonymous reviewer from Columbia University. We hope users of this edition will continue to bring to our attention any errors or parts in need of revision.

We have continued to expand the instructions for dissection because the fetal pig is the first animal many students dissect. Words of caution are given in places where students are most apt to destroy other structures inadvertently. Special attention has been given to areas where, in our experience, students have had difficulty.

We have updated sections that describe the functions of the organs being studied. Brief functional remarks reduce the laboratory time needed for instructors to explain the material and reinforce the lecture material.

We have modified some line drawings and, in some cases, have changed or added labels. White shadows have been added to all label leaders to make them easier to follow. We have kept the photomicrographs of microscopic anatomy, but omitted most of those of gross dissection. It has been our experience, confirmed by some users of the manual, that black and white photographs of dissections tend to confuse students more than help them, and it is not economically possible to use color photographs. Many of the photographs we omitted simply duplicated line drawings, but we have added line drawings in those cases where the photographs showed something different.

The Glossary has been expanded and includes simple phonetic pronunciations as well as the derivation and definition of terms. Checking terms in the glossary should help students become familiar with the basic word roots used in anatomical terminology and enable them to grasp the meanings of unfamiliar terms more easily.

Philosophical Approach

We have aimed for a flexible presentation, because these exercises are used in different types of courses; therefore, instructors need to adjust the material to meet their own needs. If instructors plan to omit particular exercises, they may wish to order the individual exercises they desire rather than the entire set. Furthermore, certain parts of individual exercises can be replaced by demonstrations, or omitted entirely. Although we are primarily addressing introductory students, we have included sufficient material to allow these exercises to meet the needs of students in some comparative and mammalian anatomy courses.

Terminology is a problem in any anatomical work because many structures have several synonyms. Human anatomists have agreed upon a set of terms, or a code, known as the *Nomina Anatomica*, although not all use it. Veterinary anatomists have adapted this to quadrupeds in a *Nomina Anatomica Veterinaria.* In both codes, the terms describe (in Greek or Latin) some aspect of the organ. It is recommended that eponyms (terms named after a person) be avoided. The human and veterinary anatomy codes differ primarily with respect to the modifying adjectives for direction: the human "superior vena cava," for example, is called the "cranial vena cava" by veterinarians. These codes are becoming the standard in anatomical writing. For the most part, we use anglicized versions of the *Nomina Anatomica Veterinaria*. The terms we favor are placed in **boldface** when first used. We avoid the use of eponyms, although we sometimes give the eponym in parentheses as a synonym if it is familiar to many people, such as **uterine tube** (Fallopian tube) and **osteon** (Haversian system). In a few cases, we use a familiar English word instead of the less familiar term from the *Nomina Anatomica Veterinaria* (e.g., "liver" rather than "hepar"); however, in such cases, the official term is given in *italics* (*hepar*) because it often forms the basis for constructing a familiar adjective (e.g., "hepatic"). A few other common synonyms are given in Roman type.

Because careful observations are essential for the acquisition of scientific knowledge and for an understanding of the generalizations that derive from an anatomical study, instructions for dissection are given in a way that will encourage students to take more than a superficial look at the organs; yet the instructions are specific enough to enable them to find structures with a minimum of assistance from the instructor. The labeled drawings are designed to help the students find structures as well as give them a record of their more important observations. Because much less laboratory time is now spent on anatomical work than formerly, students frequently do not have enough time to make their own drawings, even though the pencil is the best of aids to the eye and brain.

Authors' Preface

Each exercise is based on a natural unit of material and can be fitted into the total laboratory program where it is most appropriate. Most exercises can be completed in a standard three-hour laboratory period, although some may take a bit more time and some may take less. Certain sections can easily be omitted or replaced by demonstrations, according to the objectives of the instructor.

Acknowledgments

Fine Line Illustrations has redrawn electronically all of the art for this edition. This has allowed us to update the drawings and add new figures, but we continue to be indebted to the original artists for the great care and artistry with which they prepared the original drawings for previous editions: Judith L. Dohm, Edna Indritz Steadman, Pat Densmore, and Tom Moore. Editorial Services of New England has contributed a new and rejuvenated cover and interior design by John Svatek, and art and editorial management by Sarah Kimnach. We are also much indebted to Jodi Simpson, whose editing of these exercises has eliminated many ambiguities and errors, and to the outstanding staff at W. H. Freeman and Company who guides us through the many stages of writing and production: Erica Seifert, Project Editor; Patrick Shriner, Sponsoring Editor; Melina LaMorticella, Assistant Editor; and Maura Studley, Production Supervisor.

Warren F. Walker, Jr
Ossipee, New Hampshire, 1997

Dominique G. Homberger
Baton Rouge, Louisiana, 1997

STUDENT'S

GUIDE

I F THIS IS THE FIRST ANIMAL that you are dissecting, you should read these remarks concerning procedures and terms of direction carefully before you begin.

The purpose of a dissection is to expose you to important organs, to display their spatial relationships to surrounding organs without destroying them, and to find where the organs begin and end. If you perform a dissection well, another person should be able to examine the specimen and see clearly what the organs are, where they lie, and how far they extend.

A few cuts, or incisions, must be made at the outset with a scalpel or a pair of scissors to open the specimen, but thereafter most of the dissecting should be done with two pairs of forceps, one to hold an organ and the other to pick tissue away from it. Dissecting is done by carefully separating organs and picking away surrounding extraneous connective tissues, such as loose connective tissue, superficial fascia beneath the skin, and fat, or adipose, tissue. As you continue to dissect, you sometimes will be asked to bisect a structure; that is, to cut across its center and to reflect, or turn back, its two ends to expose an underlying organ. Do not remove any organs unless specifically directed to do so, but always pick away enough of the surrounding tissue to see all of an organ entirely and clearly. Do not be content with just a glimpse of an organ.

During the dissection, you may want to spread the legs of the specimen and tie them to the dissecting pan, but it is not necessary to do this.

The terms right and left always refer to the *specimen's* right and left sides. Depending on how the specimen is oriented, this may or may not correspond to your own right and left. Lateral refers to a direction toward the side of the body or organ in question; medial, toward the center. When a structure is described as being superficial to some part, it means that it lies over the part referred to and nearer to the body surface. Conversely, a structure that is deep to some other structure, lies beneath it and farther from the body surface.

Other terms for direction differ somewhat for a quadruped and for human beings, who stand erect (see the accompanying figure). For human beings, superior refers to the upper, or head, end of the body; inferior, to the lower parts of the body. The belly surface is anterior and the back is posterior. For quadrupeds, the underside of the body is ventral and the back is dorsal. Although the terms *anterior* and *posterior* are sometimes used to refer to directions toward the head and tail, respectively, in quadrupeds, many anatomists avoid using them in a quadruped because of their different use in a biped. Cranial describes a direction toward the head; caudal describes a direction toward the tail. Rostral describes the direction within the head toward the end of the snout. The terms *anterior* and *posterior* are used in describing some individual organs, such as an eye, where the orientation is essentially the same in human beings and quadrupeds.

The distal end of a structure is the end farthest from some point of reference, usually the origin of the structure or the midventral line of the body; the proximal end is the end nearest the point of reference.

A sagittal section is a section in the longitudinal plane of the body, passing from the middle of the back to the middle of the belly. A frontal section crosses the sagittal plane at right angles, going from the middle of the right side of the body to the middle of the left side. A transverse section crosses the longitudinal axis of the body at right angles.

The organs and structures you expose have names. Learn them so that you and your teacher can communicate effectively, but try not to be overwhelmed by them. Most terms are anglicized versions of Greek and Latin words that describe some aspect of an organ's appearance, location, structure, or function. For example, acetabulum, the socket in the hip bone for the femur, is a Latin word meaning "vinegar cup" (acetic acid has the same root); carpals, the small bones in the wrist, comes from the Greek *karpos*, meaning "wrist"; the muscle action abduction comes from two Latin words, *ab-* and *ductus*, meaning "leading away." If you look up the derivations and meanings of terms in the Glossary, you will soon become familiar with the classical roots, many of which are used repetitively, and you will learn additional anatomical terms more easily.

At the end of a laboratory period, always clean up your work area and put your specimen away in the container provided.

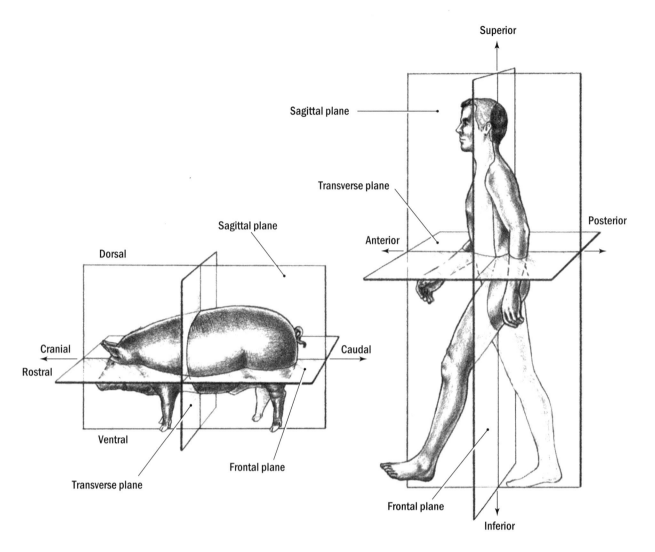

REFERENCES

The references listed below will be useful to those who wish additional information on the anatomy of the pig. The list is not exhaustive, it is simply an introduction to the literature that is available.

Biological Abstracts. Philadelphia, 1926 to date.

An essential bibliographic tool for those who wish to check the primary research literature. Two issues summarizing the biological research in the world are published monthly.

Blaney, S. P. A., and Johnson, B. *Technique for reconstituting fixed cadaveric tissues.* Anatomical Record 224: 550–551, 1989.

Chiasson, R. B., and Odlaug, O. *Laboratory Anatomy of the Fetal Pig,* 10th ed. Dubuque, IA: Wm. C. Brown Communications, Inc., 1995.

A guide for the study of the fetal pig.

Dorit, R. L., Walker, W. F., Jr., and Barnes, R. D. *Zoology.* Philadelphia: Saunders College Publishing, 1991.

Dyce, K. M., Sack, W. O., and Wensing, C. J. *Textbook of Veterinary Anatomy.* Philadelphia: W. B. Saunders Co., 1987.

A standard text in veterinary anatomy.

Fawcett, D. W. *Bloom and Fawcett, A Textbook of Histology,* 12th ed. New York: Chapman & Hall, 1994.

Although based primarily on human histology, this book is a valuable reference for all aspects of mammalian histology and cytology.

Frewein, J., Habel, R. E., and Sack, W. O. *Nomina Anatomica Veterinaria,* 4th ed. Cornell University, Ithaca, New York: World Association of Veterinary Anatomists, 1994.

Gives the official terminology for veterinary anatomy, which is becoming the standard for quadruped mammals.

Frolich, K. D., Anderson, L. M., Knutsen, A., and Flood, P. R. *Phenoxyethanol as a nontoxic substitute for formaldehyde in long-term preservation of human anatomical specimens for dissection and demonstra-*

tion purposes. Anatomical Record 208: 271–278, 1984.

Getty, R. *Sissons and Grossman The Anatomy of Domestic Animals,* 5th ed. Philadelphia: W. B. Saunders and Co., 1975.

> A useful veterinary text, now out of print.

Gilbert, S. A. *Pictorial Anatomy of the Fetal Pig,* 2nd ed. Seattle: University of Washington Press, 1966.

> A very well illustrated guide for the study of the pig. Includes many sketches of isolated organs and views of dissections not usually seen, but helpful in understanding the relationships of organs.

Junqueira, L. C., Carneiro, J., and Kelley, R. O. *Basic Histology,* 7th ed. Norwalk, CT: Appleton and Lange, 1992.

> This is an excellent, concise textbook of histology, cytology, and microanatomy, with valuable functional explanations. Although it refers mainly to human beings, it is an excellent reference for all mammalian organ systems and tissues.

Monti-Bloch, L., and Grosser, L. *Effect of putative pheromones on the electrical activity of the human vomeronasal organ and olfactory epithelium.* Journal of Steroid Biochemistry and Molecular Biology 39: 573–582, 1991.

> Adult human beings appear to have a vomeronasal organ, like all mammals, except whales and porpoises.

Nickel, R., Schummer, A., and Seiferle, E. *The Viscera of Domestic Mammals.* New York: Springer-Verlag, 1979.

Northcutt, R. G., Williams, K. L., and Barber, R. P. *Atlas of the Sheep Brain,* 2nd ed. Champaign, IL: Stipes, 1966.

Rosenberg, H. I. *How to improve the quality of the environment in the undergraduate dissection laboratory.* American Biology Teacher 54: 171–172, 1992a.

Rosenberg, H. I. *How to reduce the level of formaldehyde in the dissection laboratory.* The Morphology Teacher, No. 21, 1992b.

Sack, W. O. *Essentials of Pig Anatomy.* Ithaca, NY: Veterinary Textbooks, 1982.

> A regional guide to the anatomy of the pig.

Sack, W. O. (ed.) *Horowitz/Kramer Atlas of Musculoskeletal Anatomy of the Pig.*

> Bound with the preceding book by Sack.

Sataloff, R. T. *The human voice.* Scientific American: 108–115, 1992.

> This article presents a new theory of the mechanism of voice production by the vocal cords.

Thorpe, D. R. *The Rat and Sheep Brain.* New York: National Press Books, 1968.

> A color photographic atlas.

Walker, W. F., Jr., and Homberger, D. G. *Vertebrate Dissection,* 8th ed. Philadelphia: Saunders College Publishing, 1992.

Walker, W. F., Jr., and Homberger, D. G. *A Study of the Cat with Reference to Human Beings,* 5th ed. Philadelphia: Saunders College Publishing, 1993.

> A dissection guide to the cat, with background information on embryonic development, evolution, and function of mammalian organ systems.

Williams, P. L., et al. (eds.) *Gray's Anatomy,* 38th British ed. Edinburgh: Churchill Livingstone, 1995.

> Although based on human beings, this encyclopedic work treats all aspects of mammalian anatomy (gross, functional, microscopic, and developmental) and includes many comparative observations.

Wineski, L. E., and English, A. W. *Phenoxyethanol as a nontoxic preservative in the dissection laboratory.* Acta Anatomica 136: 155–158, 1989.

Student's Guide

External Anatomy, Skin, and Skeleton

A STUDY OF THE ANATOMY OF THE PIG *(SUS SCROFA)* IS PARTICULARLY VALUABLE, for the anatomy of the pig is closer to that of human beings than is the anatomy of other laboratory animals commonly studied. Both pigs and human beings lack much of the body hair of other mammals, both have a dentition and digestive tract adapted to omnivory, and adult pigs are closer to the size of adult human beings than cats and rats are.

Like human beings, fishes, frogs, turtles, and birds, pigs are backboned animals, a condition that places them in the **subphylum Vertebrata** of a larger group—the **phylum Chordata**. The other subphyla of the chordates include several peculiar marine forms such as sea squirts and lancelets. Chordates differ from other animal phyla, such as arthropods, in having, at least at some stage of their life cycle, a middorsal rod of turgid cells enveloped in a sheath of collagenous connective tissue that is called the **notochord**, a middorsal and tubular nerve cord, and pharyngeal pouches. The notochord is replaced in most adult vertebrates by the backbone, or vertebral column.

Among the vertebrates, pigs and human beings belong to the same **class**, the **Mammalia**. Like birds, mammals are "warm blooded," or **endothermic**. Endothermy describes the condition in which organisms use the heat generated by a relatively high level of metabolism and control body temperature by regulating the amount of heat lost at the body surface. Mammals might also be described as **homoiothermic** because they maintain their high body temperature despite ambient temperature fluctuations. Mammals, with the exception of the Australian platypus and spiny anteater, bear live young; they are **viviparous**. Newborn mammals are nursed by their mothers for some time after their birth; the mother's milk is secreted by the mammary glands.

Pigs are the domesticated descendants of European wild boars. Remains in caves in Iraq indicate that pigs were domesticated by Paleolithic people about 10,000 years ago. The wild boars found in North America are the descendants of introduced feral pigs. They are found mostly in the southeastern parts of the country, where they are called "razor backs." Wild boars are highly inquisitive, intelligent, and social animals. They are omnivorous, feeding on vegetable foods, such as roots, mushrooms, and berries, as well as animal food, such as worms, insect larvae, small mammals, and carrion.

Pigs have an even number of toes that are protected by hooves and, thus, are counted as members of the **order Artiodactyla**. Pigs are closely related to peccaries and hippopotamuses, with which they form the **suborder Suiformes**. Another artiodactyl suborder, the **suborder Ruminantia**, includes deer, cows, giraffes, and antelopes; all ruminants have multichambered stomachs and chew their cud (they ruminate). Human beings, along with lemurs, monkeys, and the great apes, belong to an entirely different group of mammals, the **order Primates**.

The gestation period for pigs is about 16 weeks; at birth, the newborn are from 30 to 35 centimeters long. Fetal pigs used for study usually range from 25 to 35 centimeters in length and vary in age from about two to three weeks to a few days before birth. The fetal pig specimens used for study are collected from slaughtered pregnant sows. A sow carries from 7 to 12 fetuses, rarely up to 16.

A. EXTERNAL FEATURES

Notice that the body of the pig, like that of most terrestrial vertebrates, consists of **head**, **neck**, **trunk** (from which two pairs of **appendages** arise), and **tail**. The head (Fig. 1-1) is marked by the typical porcine snout disk, or **rostral plate**, which bears the paired **external nostrils** (*nares*). The rostral plate is supported by an unpaired **rostral bone** (see Fig. 1-7), which occurs only in pigs. The dorsal rim of the rostral plate is hairless, thickened, and prominent. Pigs use their rostral disk for "rooting," that is, to search through leaf litter and soil for edible roots and tubers. The **mouth** is bounded by fleshy **lips**, part of which have become incorporated into the rostral plate. The external part of the ear consists of a fleshy flap, the **auricle** (*pinna*), and a short passage, the **external acoustic meatus**, which leads to the **tympanic membrane**, or ear drum. In most mammals, the external parts of the ear serve to gather and focus sound waves toward the tympanic membrane in the manner of an old-fashioned ear trumpet. The **eyes** are bounded by upper and lower **eyelids**.

Make an incision extending a few millimeters forward from the joining of upper and lower eyelids in the rostral corner of the eye, and pull the upper and lower eyelids apart. You will notice the **nictitating membrane** (*third eyelid*), which can move across the rostral part of the eyeball to distribute the tears. Human beings have a vestige of such a nictitating membrane, the semilunar fold, in the median corner of their eyes.

A mammal's trunk consists of a cranial part, the **thorax**, encased by the rib cage and housing the heart and lungs, and a caudal part, called the **abdomen** ventrally and the **lumbar region** dorsally. Liver, stomach, and intestines lie within the abdomen. In a fetal pig, the **umbilical cord** is attached to the ventral surface of the abdomen and connects the fetus with the placenta. Make a fresh cut across the umbilical cord about 1 centimeter from its attachment to the abdomen. Notice that it contains two thick-walled, round **umbilical arteries**, which carry oxygen-depleted blood and waste products from the fetus to the placenta, and one larger, thin-walled, irregularly shaped **umbilical vein**, which returns oxygen-rich blood and nutrients from the placenta to the fetus. If the arteries of your specimen have been injected, there will be colored latex in at least one of the umbilical arteries.

Between or near the umbilical arteries, you should find a small, hard cord of tissue, which is the **allantoic stalk**, a remnant of the allantois. The allantois is a fetal hol-

External Features

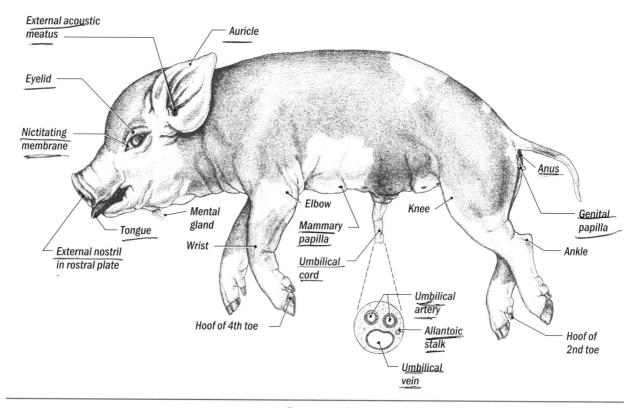

FIGURE 1-1
Lateral view of the external features of a female fetal pig. *Inset:* Cross section of the umbilical cord, enlarged.

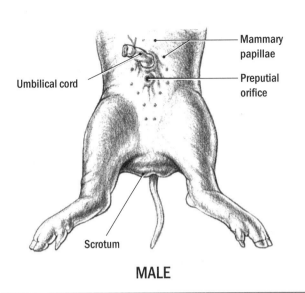

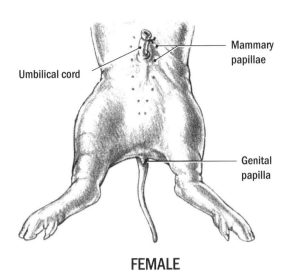

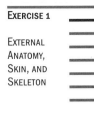

FIGURE 1-2
Ventral views of the caudal half of male and female fetal pigs, illustrating sexual dimorphism.

low organ that contributes to the formation of the placenta and will be considered later (Exercise 5). All the structures within the umbilical cord are embedded in a mucoid connective tissue known as **Wharton's jelly**.

If your specimen is a male, the **preputial orifice** of the **prepuce,** surrounding the end of the penis, is found directly caudal to the attachment of the umbilical cord on the abdomen (Fig. 1-2), and a sac of skin, the **scrotum**, which will receive the testes in a mature male, is developing caudal to the hind legs and ventral to the tail. In both sexes, the caudal orifice of the digestive tract, the **anus**, is situated directly ventral to the base of the tail. In a female specimen (Figs. 1-1 and 1-2), the common orifice of the urinary tract and vagina, which is called the **urogenital orifice**, is located directly ventral to the anus. It is bounded laterally by a pair of skin folds, the **labia**, which converge ventrally to form a spikelike **genital papilla**. This region, including the urogenital orifice and the labia, is part of the **vulva** (Exercise 5).

In both sexes, teats, or **mammary papillae**, which become part of the mammary glands in mature females, extend in paired rows along the abdominal surface. The number of pairs of mammary papillae is usually five or six, ample for nursing 6 to 12 young.

Pigs are capable of rapid sprints during escape or attack. They walk and run on the tips of their toes, a condition called **unguligrade**. Human beings, in contrast, begin a step with the entire foot and heel placed on the ground, a condition called **plantigrade**. By walking on the tips of their toes, pigs increase the effective length of their limbs and hence the length of each step. As protection, the tips of the toes are encased by hooves, which are hard keratinized structures made from material similar to that in human finger- and toenails.

The unguligrade pigs, and other hoofed mammals, such as cows and horses, are called **ungulates**. Ungulates always have some reduction in the number of their toes relative to the five toes found in most other terrestrial mammals. In pigs, the first toe, corresponding to our thumb and great toe, has been lost. Toes corresponding to our third and fourth are larger than the toes corresponding to our second and fifth. The third and fourth toes carry most of the body weight and are the main toes. Because the pig walks on its toe tips, and, because the foot is elongated, the **wrist** and **ankle** are carried well off the ground. Do not confuse these joints with the **elbow** and **knee** (see Fig. 1-1). The various segments of the limb can be determined by careful palpation of the underlying bones. They correspond to our own limb segments: **brachium** (upper arm), **antebrachium** (forearm), and **hand** in the pectoral appendage; **thigh**, **crus** (shin), and **foot** in the pelvic appendage.

External Features

Unless your specimen is unusually mature, body hair will not be conspicuous, but some relatively long hairs can be found above the eye, on the ventral three-quarters of the rostral plate of the snout, and under the chin. Those on the snout and under the chin are called **vibrissae**. Vibrissae are surrounded by touch receptors at their roots and serve as tactile sensory organs. You may notice a peeling layer of skin covering much of the body; this is the **epitrichium**, a layer of embryonic skin that is sloughed off as the hairs develop beneath it. The epitrichium, together with sloughed-off skin cells and secretions of the skin's sebaceous glands, form a cheesy **vernix caseosa** that covers the surface of the fetus and probably protects it from the surrounding amniotic fluid, which late in embryonic life contains a variety of debris from

3

the digestive and urinary tracts. The vernix is lost soon after birth.

Pigs also possess several pheromone-producing cutaneous glands, which are best developed in adults. The **mental gland** is located under the chin at the base of the chin vibrissae. **Carpal glands** are skin pockets arranged in a row of four to ten on the mediopalmar side of the forelimb dorsal to the carpus (wrist). You may be able to see these glands as pinprick-sized holes in your specimen. The secretions of both glands may serve for interindividual communication and for marking females during mating.

B. INTEGUMENT

The integument, or skin, is a body organ in its own right. It is a boundary layer, forming the interface between an animal's internal environment and the outside world. The integument protects the body against mechanical abrasions and damaging ultraviolet radiation, controls the loss of water, salts, and body heat, and generally helps to maintain a milieu within the body that is quite different from and more stable than the external environment. The integument is also a sensory organ, receiving tactile, pain, and thermal signals from the external environment.

The basic structure of the integument can be studied on a microscope slide of a vertical section through the integument of any mammal. The integument of cats is often used for study because it contains more hair than that of pigs or human beings.

The integument consists of two main layers: an outer, thin **epidermis**, and an inner, much thicker **dermis** (Fig. 1-3). Using high power on the microscope, you can see that the epidermis is a stratified squamous epithelium. An **epithelium** is a tissue in which the cells are tightly packed together. Most substances that cross an epithelium must go through the cells, which exert a regulatory role. For this reason, epithelia cover all body surfaces and line nearly all cavities. Epithelial cells rest on an inconspicuous **basal lamina** of minute intercellular fibers.

The **stratum basale** (basal layer) of the epidermis, next to the dermis, consists of cuboidal cells. (Cell boundaries are often difficult to see in animal tissues, but the size and orientation of a cell can often be inferred from its more conspicuous nucleus.) Because the cells in the stratum basale divide mitotically and generate more cells, this layer is often also called the stratum germinativum. Dividing cells can sometimes be seen.

Integument

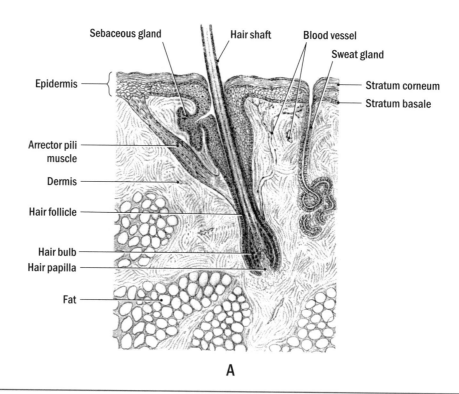

Sebaceous gland · Hair shaft · Blood vessel · Sweat gland

Epidermis

Stratum corneum
Stratum basale

Arrector pili muscle

Dermis

Hair follicle

Hair bulb
Hair papilla

Fat

A

FIGURE 1-3
Mammalian skin.
(A) Diagram of a vertical section.

As many of the new cells move toward the surface of the skin, they become flattened and filled with the horny protein **keratin**, which binds with lipids and makes the cells impermeable. Eventually the cells die and form a horny, or corneous, **stratum corneum** at the integument surface, which reduces the loss of body water. This layer loses cells at the surface as rapidly as new ones are added to it from remaining cells in the stratum basale. Intermediate layers can be recognized in thicker regions of the skin.

Dark **pigment granules** of **melanin** may be seen in the deeper cells of the epidermis. The pigment is produced by specialized cells within the deepest layer of the epidermis, or penetrate into it from the dormis. Melanin absorbs some light energy and helps to protect deeper tissues from damaging ultraviolet radiation.

The dermis consists of a dense, fibrous connective tissue. **Connective tissue** is characterized by an extensive extracellular matrix that has been secreted by the cells. The matrix consists of a viscous ground substance containing many intercellular fibers. Bundles of these fibers can be seen running in many directions. Most of the fibers consist of the protein **collagen**, which is relatively noncompliant; a few other fibers are composed of a different, more elastic protein called **elastin**.

The cellular components of fibrous connective tissue are scattered **fibroblasts**, of which only the nuclei can be seen in the integument. Bundles of striated muscle fibers, representing the insertion of integumentary muscles that move the integument, are found in the deeper parts of the dermis in many mammals. Fat, a tis-

sue that provides thermal insulation and energy storage, may also be found in the dermis as well as in the subcutaneous tissue.

A hair grows out of a **hair follicle**, which extends into the dermis from the epidermis and sometimes even into the subcutaneous tissue. (Because the plane of the section of a slide seldom parallels the axis of the hair, as it does in Fig. 1-3A, most of the hairs will be seen in oblique sections, as in Fig. 1-3B.) A hair follicle consists of a stratified epithelium that is continuous with the epidermis and is supported by connective tissue. Each follicle completely surrounds a **hair shaft**—a shaft of cornified epithelial cells that is continually generated by the mitotic division of cells in the enlarged base, or **hair bulb**, of the follicle. A **hair papilla**, containing many capillaries, protrudes into the hair bulb and nourishes the dividing cells of the developing hair. Pigment granules may be seen in the cells of the hair shaft and bulb.

In some slides, bundles of smooth muscle fibers, the **arrectores pili muscles,** can be seen. These tiny muscles are attached to the hair follicles and extend toward the integument surface. When body temperature falls, they contract and pull the hair follicles, which are normally somewhat obliquely oriented to the plane of the integument surface, into a more erect position. In animals with dense fur, this position enables the hairs to entrap more air between them and to form a more effective thermal insulating layer. These contracting muscles also depress the integument surface between the hairs, thereby forming the "goose bumps" we sometimes notice when the muscles are activated by strong emotions or cold.

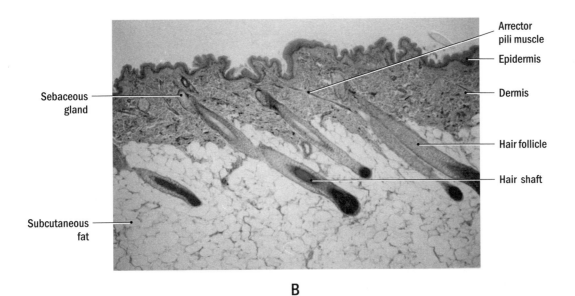

B

FIGURE 1-3 CONTINUED
Mammalian skin.
(B) Low-power photomicrograph of a vertical section.

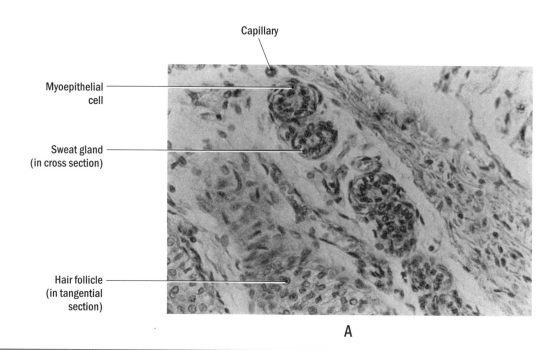

Capillary

Myoepithelial cell

Sweat gland (in cross section)

Hair follicle (in tangential section)

A

FIGURE 1-4
Mammalian skin.
(A) High-power photomicrograph of sections of a sweat gland.

Integument

Depending on the type of integument and body region you are studying, you may see integument glands, nerves, and blood vessels in the skin sections (Fig. 1-4). **Sebaceous glands** are associated with hair follicles in some regions of the body. If present in your slide, they appear as saclike clumps of lighter staining cells attached to, or lying beside, the hair follicles (Fig. 1-3B). Many of their cells are filled with oil droplets, which are discharged into the hair follicle when the cells break down. This oil conditions hair and keeps it healthy. In human beings, it also prevents excessive drying of the skin surface and keeps the skin soft and smooth.

Sweat glands are long, coiled, tubular glands. They are not common in most of the skin of heavily furred mammals, but the slides may include some. They appear in section as small clumps of cuboidal epithelial cells (Fig. 1-4A). The nuclei of the cells are relatively large and round. In examining a cross section of a glandular tubule, you may see that the cells surround a very small lumen, but shrinkage of the tissue in the course of slide preparation sometimes obliterates this cavity. Flattened nuclei peripheral to the cuboidal cells belong to elongated **myoepithelial cells**; when these cells contract, they compress the glandular tubules and assist in the discharge of sweat.

The sweat glands you usually see are **eccrine sweat glands**, which secrete some salts and excretory products; their main secretory component, water, is secreted in abundance when body temperature rises, and its evaporation removes body heat. If you examine your fingertips with a hand lens, the openings of sweat glands will appear as small pits on the fingerprint ridges. Many heavily furred mammals have only a few eccrine sweat glands and reduce body temperature by panting.

Apocrine sweat glands in the armpits and genital areas of human beings discharge their secretions into hair follicles and produce a thicker and more odoriferous secretion. In the pig, pheromone-producing apocrine sweat glands are found in the mental and carpal glands.

Blood vessels differ in appearance from sweat glands because the blood vessels have a relatively large lumen that is lined by a single layer of thin, squamous epithelial cells, whose nuclei are much smaller than those of cuboidal cells and are frequently flattened (Fig. 1-4B). The wall of a **capillary** consists of only a single layer of squamous epithelial cells, but that of a small **artery** or **vein** also contains some smooth muscle and connective tissue. Arteries have thicker walls that contain more muscle fibers than the walls of corresponding veins. The amount of blood flow through vessels near the surface of

6

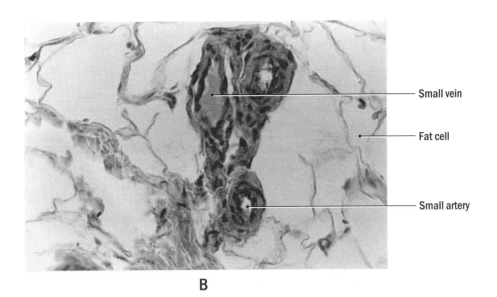

Small vein

Fat cell

Small artery

B

FIGURE 1-4 CONTINUED
Mammalian skin.
(B) High-power photomicrograph of a cross section of two small arteries
and an accompanying vein surrounded by subcutaneous fat.

Skeleton

the dermis plays an important role in regulating body temperature. Body heat is conserved by constricting these blood vessels and reducing blood flow when ambient or body temperatures are low; opposite changes occur when ambient or body temperatures are high.

C. SKELETON

The skeleton of vertebrates consists of bones and cartilages held together by connective tissue ligaments. A skeleton forms the supporting framework of the body. Beyond this, many skeletal parts—such as many bones of the skull, the vertebrae, and the ribs—encase and protect delicate internal organs; many others act as lever arms that transmit muscular forces to points of application, such as the feet; and many skeletal elements house blood-forming tissue, the red bone marrow. Study and compare mounted skeletons of a human being and of quadrupeds such as a pig or a cat (Figs. 1-5, 1-9, and 1-10). Because all are mammals, their skeletons are similar in basic features. Differences in the skull reflect differences in the shapes of the brain and sense organs, and in feeding mechanisms. Differences in the postcranial skeleton reflect differences in support and locomotion, which are largely a consequence of the type of posture of the mammal—namely, whether they are biped or quadruped.

The skeleton of vertebrates is an **endoskeleton**, because it develops within the body wall or in deeper tissues. It is not a superficial secretion on the body surface as is the exoskeleton of many invertebrates such as crustaceans and insects. The endoskeleton of vertebrates can be divided into two major parts, the somatic skeleton and the visceral skeleton. The **somatic skeleton** comprises the **axial skeleton** in the midsagittal axis of the body (most parts of the skull, vertebral column, ribs, and sternum) and an **appendicular skeleton**. The appendicular skeleton comprises the bones of the forelimb (pectoral appendage) and hindlimb (pelvic appendage), and also the shoulder and hip bones (pectoral and pelvic girdles), to which the appendicular bones attach. The **visceral skeleton** of mammals comprises the **hyoid bone**, which anchors the fleshy tongue, three **auditory ossicles** in the tympanic cavity (Exercise 6), and the **laryngeal cartilages** (Exercise 3). The visceral skeleton is best developed in fishes, in which it comprises the supporting elements for the gill and jaw apparatus, but it has become increasingly reduced in the course of evolution toward mammals.

On the basis of their embryonic development, two types of bones can be distinguished, dermal bone and cartilage replacement bone. **Dermal bone** develops directly within connective tissue in or just beneath the dermis and occurs only as part of the somatic skeleton. **Cartilage replacement bone** forms as bone replaces embryonic

7

Skeleton

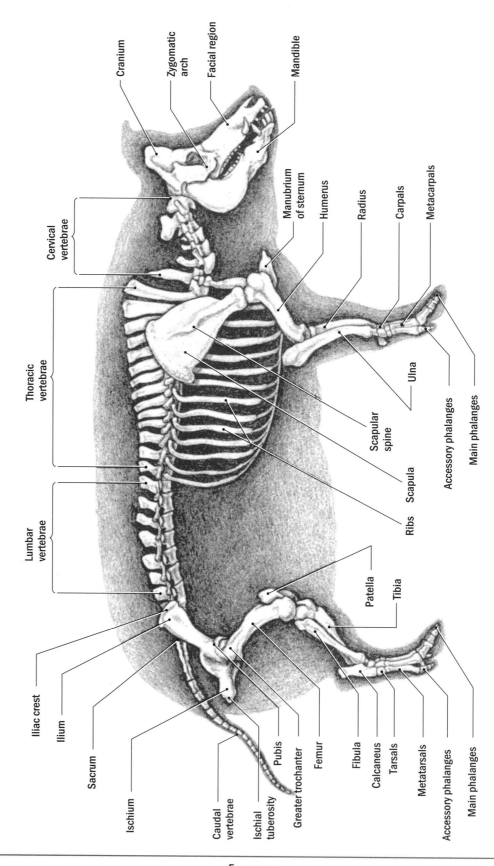

Cranium

Zygomatic arch

Facial region

Mandible

Cervical vertebrae

Manubrium of sternum

Humerus

Radius

Carpals

Metacarpals

Thoracic vertebrae

Ulna

Accessory phalanges

Main phalanges

Scapular spine

Scapula

Lumbar vertebrae

Ribs

Patella

Tibia

Iliac crest

Ilium

Sacrum

Ischium

Caudal vertebrae

Ischial tuberosity

Pubis

Greater trochanter

Femur

Fibula

Calcaneus

Tarsals

Metatarsals

Accessory phalanges

Main phalanges

FIGURE 1-5
Skeleton of a pig.

8

cartilaginous skeletal elements. This type of bone occurs as parts of both the somatic and the visceral skeletons.

C.1. Skull

Examine the general features of the skull of a human being (Fig. 1-6) and compare it with that of the pig (Fig. 1-7) or cat (Fig. 1-10). The skull consists of a **cranial region**, housing the brain and the inner and middle ear, and a **facial region**, containing the eyes and nose and forming the upper and lower **jaws**. Because of our large brain, the cranial region of a human skull is much larger and more globular than that of either a pig or a cat. The large opening at the caudal end of the cranial region is the **foramen magnum** (Fig. 1-6B), through which the spinal cord enters the skull to connect with the brain. Because of our upright posture, this foramen has rotated under the skull relative to its position in quadrupeds. The foramen magnum is

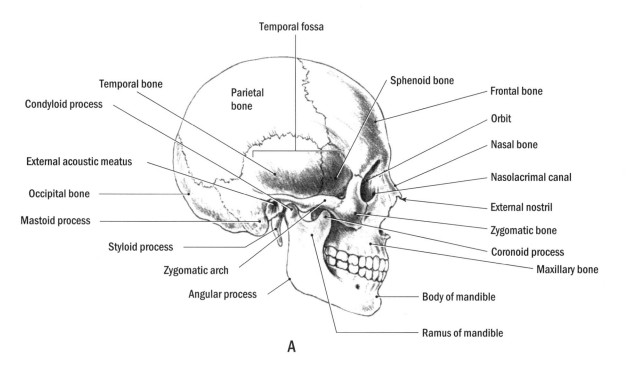

A

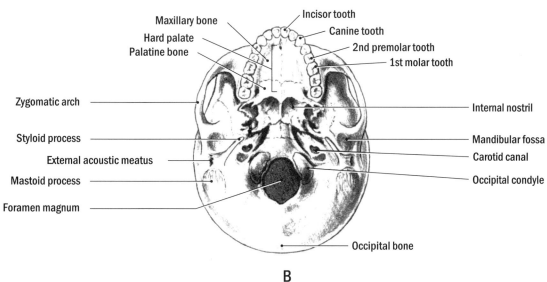

B

Skeleton

FIGURE 1-6
Human skull.
(A) Lateral view.
(B) Inferior view.

Skeleton

flanked by a pair of rounded **occipital condyles** by which the skull articulates with the first vertebra of the vertebral column. Most of the other foramina in the skull transmit nerves and blood vessels.

Observe the pair of large **orbits** that hold the eyes (Figs. 1-6A and 1-7). A depression, the **temporal fossa**, for the attachment of the temporal jaw muscles, lies caudal to each orbit. Each orbit and temporal fossa are bounded inferiorly by a bar of bone, the **zygomatic arch**, which extends from the base of each orbit to the cranium dorsal to the jaw articulation. Orbit and temporal fossa are separated by a lateral, vertical plate of bone in human beings. In cats, this dividing plate is incomplete; it is absent in the pig.

The large opening just inferior to the caudal end of the zygomatic arch is the **external acoustic meatus** (Fig. 1-6B). It leads into the **tympanic cavity**, or middle ear, from which it is separated in life by the tympanic membrane. The tympanic cavity houses the auditory ossicles, which transmit sound waves from the tympanic membrane to the oval window next to the inner ear (Exercise 6). The inner ear, which contains the receptive cells for detecting sounds and changes in body position, lies within a buttress of bone, called the **otic capsule**, within the cranial cavity. You can glimpse this capsule by looking into the cranial cavity through the foramen magnum.

The large bump of bone on the human skull caudal to the external acoustic meatus is the **mastoid process** (Fig.1-6A). Its large size in human beings correlates with the attachment of certain large neck muscles (sternomastoid, cleidomastoid, Exercise 2) that help support and balance the head atop the neck and shoulders. The mastoid process of the pig and cat are smaller than in a human being. The mastoid process of the pig is the ridge of bone ventral to the external acoustic meatus.

Quadrupeds have a large **nuchal crest** that arches across the back of the skull and receives the attachment of muscles that support the heavy head (Figs. 1-7 and 1-10). Since the human head is balanced atop the vertebral column, the human nuchal crest has shifted ventrally and is reduced in size.

A pointed **styloid process** extends ventrally from the external acoustic meatus in the human skull. In life, this process serves as the point of attachment for a ligament that suspends the hyoid bone. In the cat and pig, the hyoid bone is suspended by a chain of ossicles attached to the skull, and there is no styloid process.

A pair of **external nostrils** (*nares*) lead from the front of the face into a pair of **nasal cavities**. The passage from the rostromedial corner of each orbit into the nasal cavity is the **nasolacrimal canal**, which carries a duct through which surplus tear fluid is drained.

Examine the underside of a human skull and notice that the nasal cavities are separated from the mouth cavity by the **hard palate** (Fig. 1-6B). A pair of large **internal nostrils** (*choanae*) lie dorsal to the hard palate and connect the nasal cavities with the pharynx. The hard palate and its fleshy extension, the soft palate (Exercise 3), separate the air and food passages from each other and allow a mammal to manipulate and chew food within the mouth cavity while still breathing.

A groove near the caudal end of the zygomatic arch is the **mandibular fossa**. The condyle on the condyloid process of the lower jaw (see later) articulates here. The depth and configuration of the mandibular fossa correlates with the feeding mechanism and jaw movement of a particular species of mammal. The mandibular fossa is shallower and flatter in the omnivorous pig and ruminant herbivores than in carnivores, thus allowing more freedom of movement of the lower jaw during crushing and grinding of plant food. The relatively large **carotid canal** medial to the mandibular fossa is for the passage of the internal carotid artery, one of the main blood vessels supplying the brain. The small foramen between the carotid canal and the mandibular fossa is the auditory canal. It carries the **auditory tube**, which connects the tympanic cavity to the pharynx. Verify this by probing.

In some skull preparations, it is possible to remove the top of the skull and look into the **cranial cavity**, whose shape closely follows that of the brain. The pair of large lateral buttresses in the floor of the cranial cavity are the **otic capsules,** previously seen through the foramen magnum. Many of the individual bones that constitute the skull are shown in Figs. 1-6 and 1-7. The pig has a unique **rostral bone** that lies in front of the external nostrils and supports the rostral plate.

Examine the lower jaw, or **mandible**. It consists of a horizontal body, which bears the teeth, and a vertical ramus (Figs. 1-6A and 1-7). The ramus of the mandible articulates with the mandibular fossa by its rounded **condyloid process**. A flat **coronoid process** lies rostral to the condyloid process and serves as the attachment place of the temporal muscle, which extends downward from the temporal fossa (Exercise 2). A masseter muscle extends from the zygomatic arch to the lateral surface of the ramus and its **angular process**. You can feel these muscles bulging when you close your jaws tightly.

The teeth are deeply set in sockets in the jaw margins. Pigs have the same number and types of teeth found in the ancestors of placental mammals. On each side of the upper and lower jaws of an adult pig there are three **incisors**, one **canine**, four **premolars**, and three **molars**. The canines are longer than the other teeth; and in a boar, the upper canines curve upward to form tusks. A cat retains three incisors and the canine, but its snout is shorter than a pig's, and the number of premolars and molars is reduced. There are three premolars on each side in the upper jaw and two in the lower jaw. Only one molar per side remains in each jaw. The lower molar and the last upper premolar are enlarged and form a specialized set of

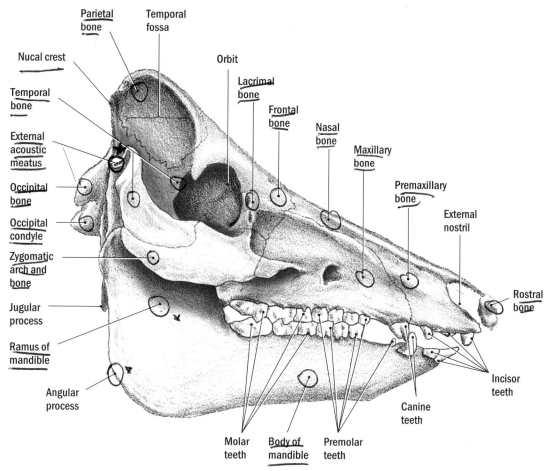

FIGURE 1-7
Lateral view of a pig skull.
(Redrawn from Getty, R. *Sisson and Grossman The Anatomy of Domestic Animals,*
5th ed. Philadelphia: W. B. Saunders Co., 1975.)

Skeleton

shearing teeth. The facial region of human beings has shortened greatly during the course of evolution, so human beings have only two incisors, one canine, two premolars, and three molars on each side of the upper and lower jaws. The canine of humans beings has become mostly incisiform, resembling an incisor, and no longer forms a long stabbing tooth as it does in the pig and cat. Its root, however, is much longer than that of an incisor.

Teeth, of course, do not grow as the jaw enlarges during postnatal development. Mammals accommodate changes in jaw size by developing two sets of teeth. A **milk** set is replaced during childhood by larger **permanent** teeth. The molar teeth are not replaced but emerge later—in human beings, at the approximate ages of 6,

12, and 18. The jaw reaches its full size by the time the last molar erupts.

C.2. Vertebrae, Ribs, and Sternum

A vertebra from near the middle of the vertebral column illustrates the basic structure of all vertebrae (Fig. 1-8). The ventral part of a vertebra is a large, solid disk of bone known as the **vertebral body**, or centrum. Dorsal to this is a neural arch, or **vertebral arch**; surrounding a **vertebral canal** through which the spinal cord passes. Cartilaginous **intervertebral disks**, which are filled with a fibrous and gelatinous material, separate successive vertebral bodies. The vertebral arches of successive vertebrae partly overlap **11** and are movably connected to one another by articular facets borne on **articular processes**. The articular facets of the pair of cranial articular processes face dorsomedially; those of the caudal pair, ventrolaterally. A dorsal **spinous process** and a pair of lateral **transverse processes** serve as sites for muscle attachment. On certain vertebrae, the transverse processes and vertebral bodies also serve as the attachment sites for ribs (see later).

The size of the vertebral bodies and the length and inclination of the spinous processes vary greatly along the

vertebral column in correlation with the varying specific stresses to which each vertebra is subjected. For example, the size of the human vertebral bodies increases steadily from the head to the pelvis because the proportion of the body weight each vertebra has to support increases.

Intervertebral foramina, through which spinal nerves pass, are located laterally between the vertebral arches. If an intervertebral disk ruptures, the distance between the bodies of the adjacent vertebrae shrinks, the intervertebral foramen becomes smaller, and the exiting spinal nerve may be pinched and become inflamed.

The vertebral column as a whole is essentially a mechanical beam that supports the head and body and transfers body weight to the appendages. No two vertebrae are exactly alike; each is modified according to the particular regional requirements for support and movement. Five vertebral regions can be recognized: cervical, thoracic, lumbar, sacral, and caudal (Fig. 1-5).

Nearly all mammals have seven **cervical vertebrae** in the neck; the first two of these, the **atlas** and the **axis**, are highly specialized to permit free movement of the head. Rocking or dorsoventral nodding movements of the head occur between the atlas and the occipital condyles on the skull; axial rotational movements occur between the atlas and axis. Lateral bending of the head occurs by lateral bending of the lower five cervical vertebrae. The transverse processes of most of the cervical vertebrae include a small embryonic rib that has fused onto them, and most of the transverse processes are perforated by a small canal through which a vertebral artery and vein lead to and from the skull.

Caudal to the cervical vertebrae are the **thoracic vertebrae**, to which ribs attach. Most ribs have a **tuberculum**, which articulates with the transverse process, and a small **head** (*capitulum*), which articulates with the vertebral body, usually between two successive vertebrae. Thus, one part of its articulation is on the caudal end of one vertebral body and the other part is on the cranial end of the next caudal vertebral body. The ribs terminate distally in flexible **costal cartilages**, most of which unite with the breast bone, or **sternum**, to form a thoracic rib cage that contributes to body support, protects the heart and lungs, and helps ventilate the lungs by its movements. The most cranial segment of the sternum, known as the **manubrium**, articulates with the clavicle in some mammalian species (see later). The number of thoracic vertebrae varies among mammalian species; human beings have 12, cats 13, and pigs 14-15.

Lumbar vertebrae lie caudal to the thoracic vertebrae and have large transverse processes with which embryonic ribs have fused. Human beings have five lumbar vertebrae, pigs six or seven and cats seven.

Caudal to the lumbar vertebrae are the **sacral vertebrae**, which are fused together to form a firm **sacrum** for the attachment of the pelvic girdle. Most mammals have

Skeleton

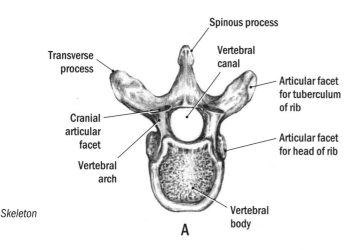

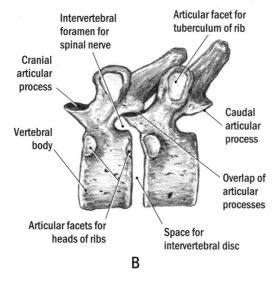

FIGURE 1-8
Human thoracic vertebrae.
(A) Superior view.
(B) Lateral view of two adjacent vertebrae.
(Redrawn from Williams et al. editors). *Gray's Anatomy*,
38th British ed. Edinburgh: Churchill Livingstone, 1995.)

three or four sacral vertebrae, but human beings have five, the greater number being correlated with the need to support all parts of the body superior to the pelvis by the pelvis and sacrum in this bipedal species. **Caudal vertebrae** supporting a tail follow the sacrum. In human beings, there are only three to five small caudal vertebrae fused together to form a **coccyx**, to which certain anal muscles attach.

C.3. Appendicular Skeleton

The major element of the pectoral girdle is a large, blade-shaped, triangular **scapula** (Figs. 1-5, 1-9, and 1-10). It is connected to the vertebral column by muscles and connective tissue; in most mammals, it has no direct bony articulation to the vertebral column. The scapula bears a long ridge on its lateral surface, the **scapular spine**, to which muscles attach; and the scapular spine terminates near the shoulder joint in a process known as the **acromion**. At the shoulder joint, the scapula forms a socket, the **glenoid fossa**, which receives the ball-shaped proximal articular end of the humerus (the bone of the upper arm).

Cranial and medial to the glenoid fossa of the human being is a beak-shaped process, the **coracoid process**, which is a remnant of a much larger bone in the pectoral girdle of nonmammalian terrestrial vertebrates. Certain arm muscles (coracobrachialis, part of the biceps, Exercise 2) attach here.

Human beings have a well-developed collarbone, or **clavicle**, extending from the acromion on the scapular spine to the manubrium of the sternum. It thus connects the scapula and shoulder joint with the sternum and the rib cage. The human clavicle braces the shoulder joint and helps to stabilize its position, but it is lost or reduced in many quadrupeds. The cat has a small one imbedded in shoulder muscles (Fig. 1-10), and it is reduced to a clavicular tendon in the pig (Exercise 2). Reduction or loss of the clavicle is an adaptation for running. Absence of the clavicle allows greater freedom of movement to the scapula and it can participate in the swing of the leg, thus increasing stride length.

The bones of the pectoral appendage are the **humerus** in the upper arm, the **radius** and **ulna** in the forearm, **carpals** in the wrist, **metacarpals** in the palm, and **phalanges** supporting the free parts of the fingers and toes, or digits. The distal end of the radius is adjacent to the first finger because it lies on the median side of the hand when the hand is in the quadruped position with the palm facing back or the sole facing the ground. The proximal end of the ulna has a process, the **olecranon**, that extends beyond the elbow joint and serves as the place of insertion for the triceps muscle (Exercise 2).

Each half of the pelvic girdle consists of three bones fused together. All three meet to form a socket, the **acetabulum**, which receives the proximal, ball-shaped head of

the femur (the proximal leg bone). The **ilium** extends dorsally and attaches directly to the sacrum. The **pubis** lies ventral and cranial to the acetabulum, and the **ischium** ventral and caudal to the acetabulum. The ilium bears a prominent **iliac crest**, which is located on the craniodorsal part of the bone, to which muscles attach. The ischium bears an **ischial tuberosity**, also for muscle attachment (Figs. 1-5 and 1-10, Exercise 2).

In adult mammals, the three bones fuse together, forming the hip bone, **os coxae**. The ischium and pubis

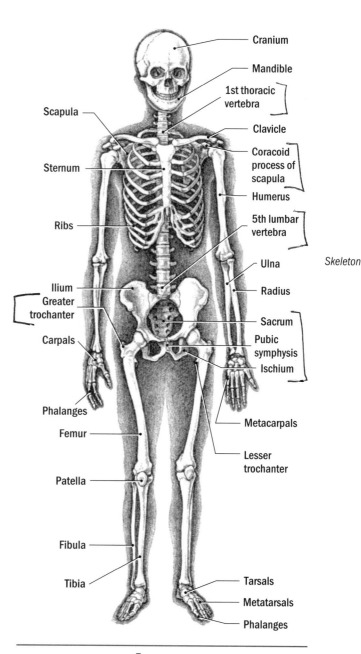

Skeleton

Labels: Cranium, Mandible, 1st thoracic vertebra, Clavicle, Coracoid process of scapula, Humerus, 5th lumbar vertebra, Ulna, Radius, Sacrum, Pubic symphysis, Ischium, Metacarpals, Lesser trochanter, Tarsals, Metatarsals, Phalanges, Scapula, Sternum, Ribs, Ilium, Greater trochanter, Carpals, Phalanges, Femur, Patella, Fibula, Tibia

FIGURE 1-9
Anterior view of the skeleton of a human being.

13

Skeleton

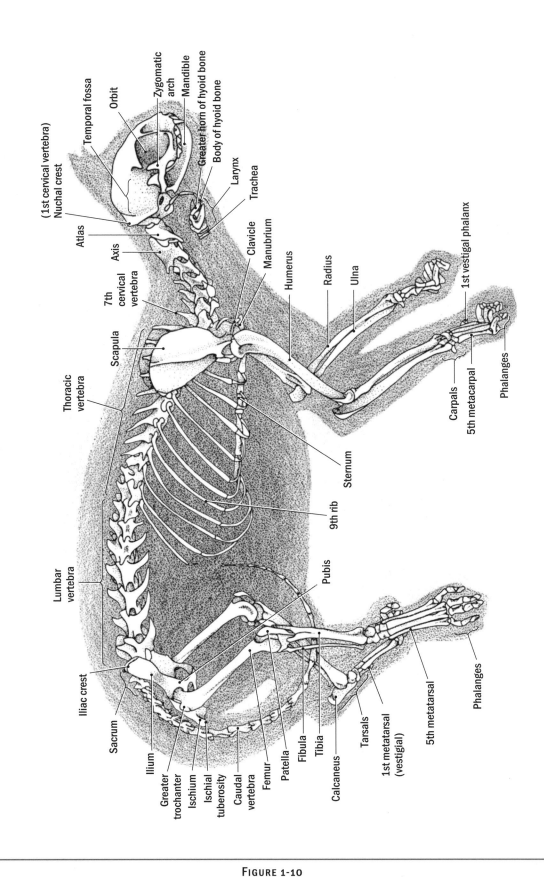

FIGURE 1-10
Lateral view of the skeleton of a cat.
(Redrawn from Walker, W. F., Jr. and Homberger, D. G. *A Study of the Cat with Reference to Human Beings*,
5th ed. Philadelphia: Saunders College Publishing, 1993.)

are separated from each other by a large opening, the **obturator foramen**, which in life is closed by a sheet of connective tissue from which certain pelvic muscles arise. The two pubic bones, one from each side of the body, join together to form the **pubic symphysis**. The paired ossa coxae and the sacrum enclose the **pelvic canal** through which the digestive, urinary, and reproductive tracts pass.

The proportions of the three bones in the human pelvis are very different from those in the quadruped pelvis. The human ilium, in particular, is relatively short but very broad and flaring. This shape provides a large surface area for the attachment of the buttock (gluteal) muscles (Exercise 2), which help stabilize the trunk over the legs. The bowl-shaped human pelvis, together with the sacrum, also supports the abdominal organs.

The bones of the pelvic appendage are the **femur** in the thigh, the stout **tibia** and thin **fibula** in the shank, **tarsals** in the ankle, **metatarsals** in the sole, and phalanges in the free part of the toes or digits. The femur bears several bumps for muscle attachment on its proximal end. The largest bump, on its lateral surface is the **greater trochanter** (Figs. 1-5 and 1-10). The particularly long tarsal bone that extends caudally from the foot is the **calcaneus**. Powerful shank muscles attach here (Exercise 2).

Limb and foot structures are closely adapted to the use of the limb for locomotion. As we have discussed, pigs are unguligrade and walk on their toe tips. Their first toe is lost, the weight-bearing main toes (three and four) are enlarged, and the foot is elongated, especially the metacarpals and metatarsals. During locomotion, the limbs swing back and forth like pendulums. The front leg of the pig is not used for manipulating food, so it does not need to be rotated. The radius does not move in relation to the ulna, and the two bones are tightly linked together. (They are fused in cows.)

Cats are good runners, but they use their front leg also to manipulate food and groom themselves. Their feet are elongated, and they walk by placing the entire length of the free parts of the digits on the ground, with their wrist and ankle lifted above the ground, a position called **digitigrade**. The first digit is reduced to a vestige in the hind foot but retained, though smaller, in the front foot. The radius is able to rotate about the ulna so that the foot may be **prone** (sole toward the ground) or **supine** (sole facing upward). The terminal phalanges, which bear the claws, are hinged on the next, penultimate phalanges in such a way that the claws can be retracted or extended.

The human arm and hand are used in grasping, not locomotion. Much rotation is possible, and the thumb can be opposed against the other digits. Humans beings are bipeds. The leg is large and powerful, and the foot is adapted to carry the entire weight of the body. As noted earlier, we have a plantigrade walk, with the entire foot initially placed on the ground. The great toe, with which we push off the ground, is enlarged.

If a skeleton of a young mammal is available, notice the conspicuous transverse cleft near each end of the long limb bones. In life, a cartilaginous **epiphyseal plate** is located here. Its growth and replacement by bone is responsible for the growth in length of the bones. At puberty, the cartilage of the epiphyseal plate is completely replaced by bone, and growth stops.

D. SKELETAL MATERIALS

D.1. Cartilage

Cartilage and bone are modified connective tissues that form the skeleton of vertebrates. Both resist compression well, but bone is more resistant to tension and twisting. Cartilage is more compliant and can grow from within through expansion without the complex remodeling and replacement required in bone. Examine a slide of cartilage. It consists of cells, **chondrocytes**, that lie in spaces, the **lacunae**, in an **extracellular matrix** that is secreted by the chondrocytes (Fig. 1-11). The matrix consists of **collagen** fibers that are masked in a ground substance containing a great deal of bound water. Cartilage is not supplied by blood vessels and receives nutrients and oxygen by diffusion from its surface. This lack of a blood supply limits the size of cartilaginous elements.

Translucent, glasslike cartilage is found, among other places, on the ends of the long bones of the limbs and is know as **hyaline cartilage** (Fig. 1-11A). Visible, elastic fibers consisting of the protein **elastin** are abundant in **elastic cartilage** (Fig. 1-11B), which is found, for example, in the auricle of the ear. The smoothness and compliance of cartilage is a result of its high water content.

Cartilage grows internally by the division of the chondrocytes, which then separate and produce more matrix. It also grows peripherally through the transformation of fibroblasts into chondrocytes in the connective tissue **perichondrium**, which surrounds cartilage. These growth processes allow rapid expansion; therefore, cartilage forms much of the skeleton of embryonic vertebrates. Cartilage persists in adults wherever a smooth, firm, and compliant tissue is needed, such as on the articular surfaces of the bones and in the costal cartilages.

Skeletal Materials

D.2. Bone

The major skeletal material of adult vertebrates is bone, which is a highly vascularized and mineralized, dense connective tissue. Bone is particularly strong because its mineral component resists compression and its fibrous component provides some compliance and resists tension and twisting. Bone, like cartilage and fiberglass, is a composite material made of several substances with different properties. The spread of cracks that may start in a

15

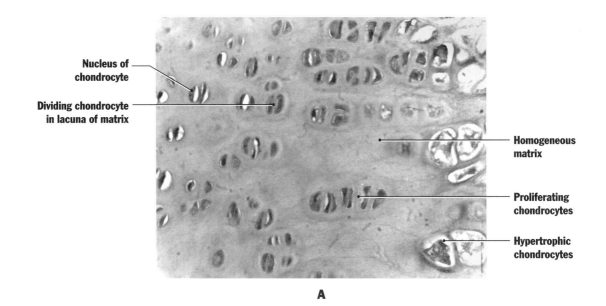

Nucleus of chondrocyte

Dividing chondrocyte in lacuna of matrix

Homogeneous matrix

Proliferating chondrocytes

Hypertrophic chondrocytes

A

Skeletal Materials

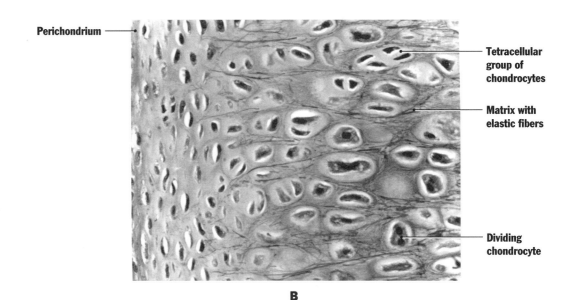

Perichondrium

Tetracellular group of chondrocytes

Matrix with elastic fibers

Dividing chondrocyte

B

FIGURE 1-11
Photomicrographs of cartilage taken at a magnification of 450×.
During slide preparation, the cartilage cells shrink, leaving a space between them and the surrounding matrix.
(A) Hyaline cartilage from a long bone of a kitten. Articular cartilage is on the left, and merges with the cartilage of an epiphyseal plate on the right. An epiphyseal plate is a plate of cartilage located near the end of a long bone of a young mammal, where growth in length of the bone occurs. Cartilage cells in the epiphyseal plate proliferate rapidly and begin to hypertrophy and show other changes as the cartilage is replaced by bone.
(B) Elastic cartilage from the external acoustic meatus of a dog.

composite material is usually blunted at the interface between two different components. Living bones are covered by a layer of connective tissue called the **periosteum** (Fig. 1-12).

Compact bone forms the periphery of the long bones of a limb, and **spongy bone** lies deep to this near the ends of the bone. Marrow fills the large cavities of bone. Bone, like cartilage, consists primarily of an extracellular matrix, but the matrix is composed of hard inorganic salts (mostly crystals of a calcium and phosphorus compound known as **hydroxyapatite**). These crystals are bound to fibrous proteins, chiefly collagen, in the matrix.

16

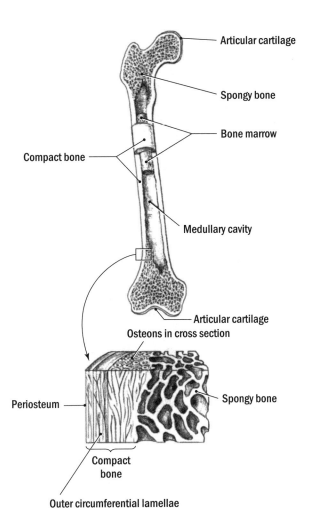

FIGURE 1-12
Structure of the femur, a representative bone.

Examine a slide of a cross section of compact bone (Fig. 1-13). Most of the bony material is deposited in concentric rings around tiny blood vessels that tend to parallel the longitudinal axis of the bone. Each of these columnar units, which appear circular in cross section, is an **osteon**, or Haversian system. In the center of each is a **central canal** containing, in life, one or more blood vessels (usually capillaries and venules) embedded in loose connective tissue. Around this canal are concentric layers of bone matrix, the **concentric lamellae**. Between the lamellae are rows of dark dots, the **lacunae**, which in life contain the cellular elements of the bone (**osteocytes**). Minute **canaliculi** extend more or less radially from the lacunae and contain processes of the osteocytes. Bone is a dynamic tissue, and much of it is reabsorbed and rebuilt continuously. Parts of former osteons (**interstitial lamellae**) can be seen between the ones formed more recently. **Outer circumferential lamellae** lie at the bone surface.

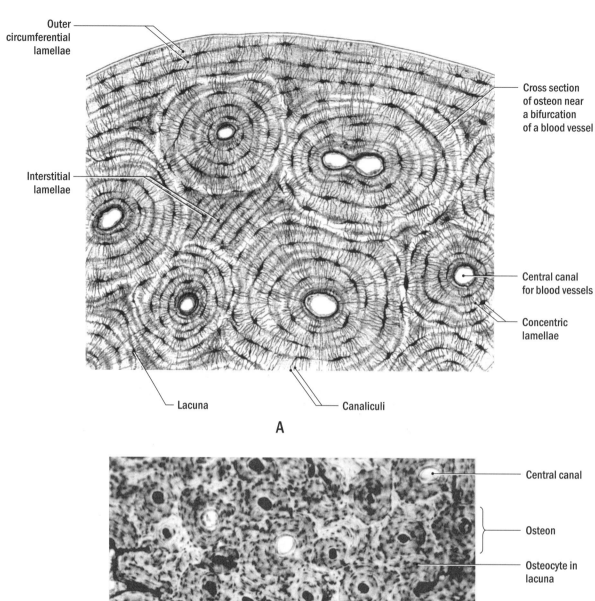

Outer circumferential lamellae

Interstitial lamellae

Cross section of osteon near a bifurcation of a blood vessel

Central canal for blood vessels

Concentric lamellae

Lacuna

Canaliculi

A

Skeletal Materials

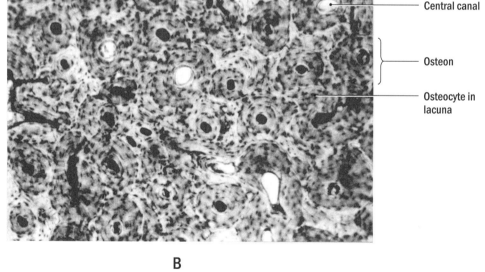

Central canal

Osteon

Osteocyte in lacuna

B

FIGURE 1-13
Microscopic structure of a mammalian long bone.
(A) Diagram of a group of osteons as seen in a transverse section of bone.
(B) Low-power photomicrograph of a transverse section through dried bone.

Muscles

M USCLES CONSTITUTE THE LARGEST ORGAN SYSTEM in the body, as measured by volume. They are present in all parts of the body and, ultimately, they are responsible for all movements of and within the body, such as the movements of the bony skeleton, the progress of food through the digestive tract, the flow of blood through the circulatory system, the release of urine from the urinary bladder, and the secretion of saliva from the salivary glands. Some muscles are inconspicuous because they are part of the skin or the walls of inner organs. Other muscles are very conspicuous and organlike. For example, the heart consists almost completely of a special type of muscle, and the muscles attaching to the skeleton are individually identifiable. We can classify muscles into two or three groups, depending on their position within the body, their innervation, and their histological structure (see Table 2-1 and Section K entitled Muscle Tissue).

Table 2-1
Classification of Muscles

GENERAL LOCATION	INNERVATION	HISTOLOGICAL TYPE	SPECIFIC LOCATION	RELATIONSHIP TO GENERAL BODY PLAN
Skeleton	Voluntary	Striated	Most skeletal muscles	Somatic
			Some larynx, jaw and shoulder muscles	Branchiomeric (special visceral)
Inner organs and hair follicles	Involuntary	Smooth	Most inner organs and hair follicles	Visceral (general visceral)
		Cardiac	Heart	

We will dissect only the major skeletal muscles. Most of these muscles are **somatic muscles** and are associated with the somatic part of the skeleton. Some muscles of the larynx, jaw, and shoulder are, technically speaking, **visceral muscles**. These are called **branchiomeric muscles** because they are derived from muscles that were associated with the gill (branchial) apparatus of ancestral fishes.

The muscles of the fetal pig are representative of those of adult mammals, but fetal muscles are thinner and softer than those of an adult individual, and they tear easily (see Section C entitled Dissection Technique for Muscles).

A. REMOVING THE SKIN

Removing the skin from a fetal pig is an exacting task that needs to be done with great care if you want to be able to identify the body muscles.

To remove the skin from one side of the body, make an incision through the skin slightly to the side of the mid-dorsal line above the shoulders. You can palpate the mid-dorsal line by feeling the middorsal row of spinous processes of the vertebral column. While you are making the incision, pull upon the cut edge with forceps to avoid cutting underlying muscles. The most superficial muscle you will see is the trapezius (see later, Fig. 2-3), which will provide you with a landmark for the proper level between the skin that you have to remove and the surface of the body musculature and connective tissue sheets that you must not injure. Make sure to maintain this level, even as you move beyond the trapezius muscle to the superficial sheets of connective tissue to remove the skin from the body.

Extend the longitudinal incision to both the back of the head and the base of the tail. As you reflect the skin, make additional incisions (use a pair of scissors to cut through the separated skin) that extend from the middorsal line ventrally to the left ear and across the cheek to the chin, around the tail and genital area, and down the lateral surfaces of the legs. Carefully peel off the skin from the body, leaving the skin on the head, feet, tail, and genital area, including the scrotum and penis if you have a male specimen, and around the umbilical cord.

The loose connective tissue beneath the skin is known as the **superficial fascia**. It can easily be torn by using two pairs of forceps to pull it apart. Do not injure deeper muscular tissue, which looks a bit darker and consists of parallel strands of muscle fibers. If you wonder whether you are dealing with muscle or connective tissue, lift the tissue sheet in question and hold it against a light. Connective tissue is usually translucent, whereas muscle tissue is not. The tougher **deep fascia**, which encases some muscles, should be left on the body for the time being.

As you remove the skin, cut the blood vessels and nerves that emerge from the trunk musculature at certain intervals and supply the fascias and the skin. This pattern of intervals is an indication of the basic segmental organization of the vertebrate body, which is much more obvious in fishes than in mammals.

Over the thorax, a broad muscle (the latissimus dorsi, see Fig. 2-3) tends to stick tightly to the skin; carefully separate it from the skin and leave it on the body. As you approach the elbow region, you will notice connective tissue and muscle fibers arising caudally from the armpit and extending to the lateral flank skin, to which it is tightly attached. These muscle fibers are parts of the **cutaneus trunci**, an extensive but sometimes inconspicuous sheet of muscle. The cutaneus trunci fans out from its origin on the pectoral muscles, in particular over the caudolateral

border of the pectoralis profundus muscle (see Figs. 2-3 and 2-4), and along the midventral line to run caudally and caudodorsally to insert into the skin of the trunk, which it can move. The cutaneus trunci muscle is present in most mammals but is absent in human beings. With a sharp scalpel, cut across the muscle fibers that are stuck to the inner surface of the flank skin about 4 centimeters from their origin on the pectoral muscles and the midventral line. Separate the cutaneus trunci fibers from the skin near their origin and remove the rest of the muscle with the skin. The cutaneus trunci is a somatic muscle.

As you proceed to remove the skin from the neck, you will discover another cutaneous muscle, the **platysma**, which covers the ventral and lateral surfaces of the neck. It arises from the shoulder region and sends its longitudinally oriented muscle fiber bundles onto the face, where it breaks up into numerous **facial muscles** associated with the eyelids, lips, nose, and auricles of the ears. The platysma and facial muscles are branchiomeric in origin. Some of these muscles make the cheeks and lips of mammals fleshy and give mammals their unique ability to suckle during infancy.

The platysma and facial muscles are very tightly bound to both the skin and the body muscles. Separate them from the skin, leave them on the body until the skin is removed, and then take them off very carefully.

A thin, but extensive, salivary gland, the **parotid gland** (Fig. 3-1, Exercise 3), lies on the side of the neck deep to the platysma. It is often stuck to the inner surface of the platysma and must be separated very carefully. It may be studied now (see Exercise 3) or removed with the platysma and examined later on the other side of the body. The platysma and parotid gland lie superficial to a large vein (the **external jugular vein**) and its tributaries. Do not destroy them.

B. MUSCLE TERMINOLOGY AND MUSCLE FUNCTION

An individual skeletal muscle consists of many **muscle fibers**, or muscle cells. Each muscle fiber holds a multitude of myofibrils, which contain the contractile elements of muscles (see Section K). Muscle fibers are held together by connective tissue to form **muscle fiber bundles** (fascicles), which again are held together by connective tissue to form muscle bellies (Fig. 2-1). The smallest units discernible with the naked eye are muscle fiber bundles. Near the attachment of a muscle to a bone, each muscle fiber attaches to collagen fibers through a **myotendinous junction**. The collagen fibers are held together by connective tissue to form a cordlike **tendon** for a spindle-shaped muscle or a sheetlike **aponeurosis** for a sheetlike muscle. The tendons and aponeuroses attach to a bone by interweaving

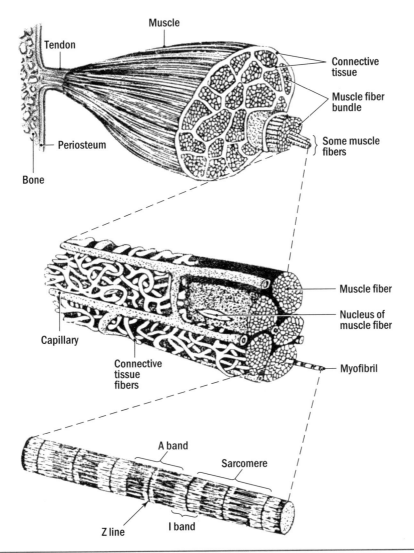

Muscle Terminology and Muscle Function

FIGURE 2-1

Diagram of the structure of a skeletal muscle, its attachments to bones, and its microscopic composition. (Redrawn from Walker, W. F., Jr., and Homberger, D. G. *Vertebrate Dissection*, 8th ed. Philadelphia: Saunders College Publishing, 1992. After Dorit, R. L., Walker, W. F., Jr., and Barnes, R. D. *Zoology*. Philadelphia: Saunders College Publishing, 1991. And Fawcett, D. W. *Bloom and Fawcett, A Textbook of Histology*, 12th ed. New York: Chapman & Hall, 1994.)

their collagen fibers with the connective tissue of the periosteum enveloping the bone and by sending collagen fibers into the bone substance for anchoring.

Although a contracting muscle exerts an equal force on both skeletal elements to which it attaches, by convention, its attachment to the more stationary and proximal skeletal element is called its **origin** and the attachment to the more mobile and distal skeletal element is called its **insertion**. An individual muscle may have more than one origin and one insertion. It is not always easy to determine which attachments of a particular muscle are its origin or insertion. In such cases, it may be better to use the neutral term of **attachment** for all ends of a muscle.

Normal movements of the body are rarely generated by single muscles acting alone. Instead, several muscles usually act together as a group, that is, **synergistically**. Bending or flexing just the distal ends of your fingers, for example, requires not only the contraction of muscles on the palmar side of the forearm, but also the contraction of muscles on the opposite side to stabilize the proximal finger segments and prevent them from bending.

Contracting muscles generate forces and thereby create tension (**isometric contraction**) or shorten (**isotonic contraction**). Because muscles cannot lengthen actively, an opposing, or **antagonistic**, force is required to restore the contracted muscles to their resting length. Usually it is

21

another muscle or muscle group that provides such an antagonistic force. In general, one muscle group (agonist muscles) moves a skeletal element in one direction, and another muscle group (antagonist muscles) moves it in the opposite direction. Muscles can be sorted into mutually antagonistic groups that perform opposite functions. The following terms define the more common antagonistic actions.

Flexion and **extension**. Flexion is a bending; that is, it is the movement of a distal body part toward a more proximal part, such as that occurring when the forearm is moved toward the upper arm. Extension is the opposite action (Fig. 2-2).

Protraction and **retraction**. Protraction is a forward movement of an entire limb at the shoulder or hip joints; although the term *flexion* is sometimes used for this movement, the term *protraction* is more accurate for a quadruped. Retraction is a backward movement of an entire limb.

Adduction and **abduction**. Adduction is a movement of a body part toward a point of reference; abduction is a movement away from a point of reference. The point of reference for adduction and abduction of a limb is the midventral line of the body.

Many students are at first overwhelmed by the large number of unfamiliar names for muscles; but many of these names can be very helpful, because they may be

descriptive of the attachments (e.g., sternomastoid), shape (e.g., trapezius), number of divisions (e.g., triceps), or function (e.g., tensor fasciae latae) of muscles.

C. DISSECTION TECHNIQUE FOR MUSCLES

Only the more conspicuous muscles can be closely observed in an exercise of this scope. Before attempting to identify muscles, carefully clean their surfaces by picking away loose connective tissue and fat. Then identify the borders of individual muscles by carefully watching the orientation of the muscle fiber bundles, because it often varies in adjacent muscles and depends on their attachments. Once you have identified the borders of muscles, you can separate the individual muscles from each other by using two pairs of forceps to break the connective tissue that binds the muscles together. If you proceed correctly with your dissections, the surfaces of the muscles will remain smooth and intact.

If it is necessary to transect a superficial muscle to reveal a deeper one, you may first want to ascertain the orientation of all its muscle fiber bundles and free all its borders from connective tissue. Then lift one border with a

*Dissection
Technique for
Muscles*

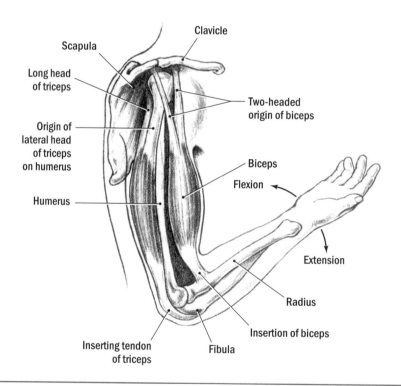

FIGURE 2-2
Diagram of an antagonistic pair of skeletal muscles and their actions in the human arm.

pair of forceps and cut with a pair of scissors (using the blade with the blunt tip as the lower blade) as far as you can through the center of the muscle perpendicularly to its muscle fiber direction. If necessary, free the muscle from the underlying muscles before you extend the cut through the entire muscle and reflect the two halves of the bisected muscle. Bisect only one muscle at a time. This procedure will facilitate reconstructing superficial muscles when you wish to review them.

The muscles of the pig at the fetal stage of development are not as well developed as they are in young or adult specimens. Fetal muscles are small, thin, usually very pale, and easily torn. In contrast, the connective tissue that surrounds and connects the individual muscles is well developed and often tougher and stronger than fetal muscle tissue. Nerves leading to the muscles often form extremely strong threads that cannot be torn and must be bisected with a pair of scissors. These conditions conspire to make the dissection of muscles in the fetal pig a challenging, though rewarding exercise. The dissection is considerably facilitated by using watchmaker's forceps as an additional dissection tool and by performing the dissection under a magnifying lamp or a dissection microscope. If possible, muscles should be studied in a fetus in the last stage of prenatal development.

D. MUSCLES OF THE SHOULDER

The first muscles to be considered are those that extend from the trunk to the pectoral girdle, as well as those that extend from the trunk and pectoral girdle to the proximal part of the humerus. Collectively, they move the shoulder and arm as a whole, or they move the trunk relative to the foot when the front foot remains fixed on the ground.

D.1. Superficial Muscles of the Shoulder

Clean up the lateral surface of the shoulder, neck, and upper arm as well as the ventral surface of the chest as directed above before attempting to identify the individual muscles. Reference to Figs. 2-3 and 2-4 will help you find the separations between the muscles.

The most superficial muscle is the thin, triangular **trapezius**, which arises from the occipital bone of the skull and from the dorsal surface of the neck and thorax. Its muscle fibers converge to a broad aponeurosis to insert on the scapular spine, which you can palpate. The muscle helps to hold the scapula in place and to move it forward and backward. Most muscles of the shoulder are somatic, but the trapezius is branchiomeric because it has evolved from gill arch muscles in ancestral fishes, which acquired a secondary attachment to the pectoral girdle.

The large, triangular muscle that lies caudal to the brachium (upper arm) and fans out over the back is the **latissimus dorsi**. It arises from the dorsal surface of the thorax deep to the caudal part of the trapezius and from a tough sheet of deep fascia on the back (the **thoracolumbar fascia**). It passes into the armpit to insert on the proximal part of the humerus, deep to a large arm muscle, the triceps (see Section E). The latissimus dorsi is a major retractor muscle of the arm.

Cranial to the trapezius and extending from the back of the neck and head to its insertion on the proximal part of the humerus is the complex **brachiocephalicus**. It helps to protract the arm and consists of three band-shaped portions in the pig (Figs. 2-3 and 2-4). The most superficial portion, the **cleidooccipitalis**, originates from the nuchal crest of the skull just ventral to the origin of the trapezius and follows the dorsal half of the cranial border of the trapezius toward the shoulder. The **cleidomastoid** arises ventrally on the caudal end of the zygomatic arch of the skull and follows at first the cranioventral border of the cleidooccipitalis. Toward the shoulder, it turns deep to the cleidooccipitalis and fuses with it at the level of the shoulder joint. In an adult pig, both muscles insert together on the **clavicular tendon**, which is not yet fully developed at the fetal stage. The third part of the brachiocephalicus, the **cleidobrachialis**, arises from the fusion point of the cleidooccipitalis and cleidomastoid (or from the clavicular tendon in an adult pig) and inserts on the proximal end of the humerus. It covers the cranial aspect of the shoulder joint.

Muscles of the Shoulder

The cleidooccipitalis and cleidomastoid develop embryonically as parts of the trapezius and are, therefore, branchiomeric. The cleidobrachialis, in contrast, develops embryonically as a part of the deltoid (see later) and is, therefore, somatic. In human beings, the clavicle is fully ossified, and there is no brachiocephalicus as such; the cleidomastoid connects the skull to the clavicle, and the cleidobrachialis arises from the clavicle independently and laterally from the cleidomastoid.

Separate the cleidooccipitalis, cleidomastoid, and trapezius from one another. If you push the borders of the cleidooccipitalis and trapezius apart, you will discover a narrow, band-shaped **omotransversarius**, which arises from the prominent wing of the atlas deep to the previous muscles. As it runs toward the scapula, it emerges from under the cleidooccipitalis. It follows the ventral border of the trapezius and inserts on the scapular spine just ventral to the insertion of the trapezius. The omotransversarius helps to protract the scapula. Human beings do not have this muscle.

The **deltoid** arises ventral to the insertion of the trapezius from the scapular spine and from a fascia over deeper muscles on the lateral surface of the scapula (the infraspinatus and supraspinatus, see later). It extends ventrally to insert on the fascia covering the lateral surface of

23

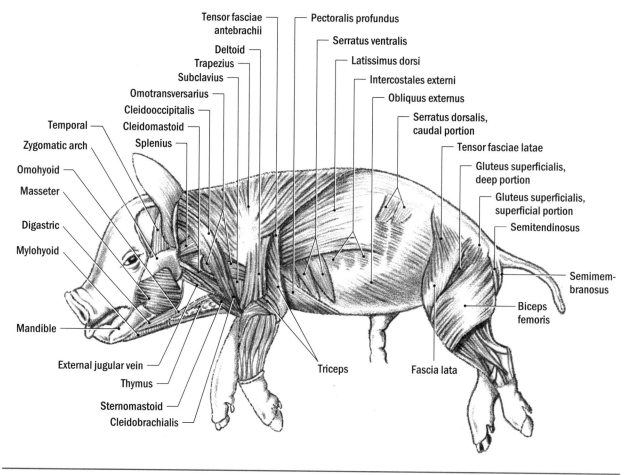

FIGURE 2-3
Lateral view of the superficial muscles of a fetal pig after removal of the cutaneous muscles and the parotid gland.

*Muscles of
the Shoulder*

the upper arm on the humerus next to the insertion of the cleidobrachialis. It helps to protract the upper arm.

Free the omotransversarius from the trapezius and any underlying tissue and bisect it. Lift the cranial border of the trapezius and free it from underlying tissue and muscles. With a pair of scissors, bisect the freed muscular part of the trapezius as far as you can, proceed with freeing the rest of the muscle, bisect it, and so on until you have reached the caudal border and have bisected the entire trapezius. Reflect both halves of the trapezius and omotransversarius.

The ventral surface of the chest is covered by the large, triangular **pectoralis**, which consists of three parts (Fig. 2-4). The more superficial **pectoralis superficialis** (which is homologous to the pectoralis major of human beings) arises from the cranial half of the sternum and inserts on the humerus medially along the insertion of the cleidobrachialis and on the fascia covering the proximal part of the medial surface of the forearm. It adducts and retracts the arm.

The deeper **pectoralis profundus** (which is homologous to the pectoralis minor of human beings) arises

from the sternum caudal to the origin of the pectoralis superficialis. Its muscle fibers converge and extend craniolaterally deep to those of the pectoralis superficialis. Trace them by cutting through the pectoralis superficialis after you have separated it from the underlying pectoralis profundus. The pectoralis profundus inserts on the proximal part of the humerus.

The third muscle of the pectoral group is the **subclavius**; it has no counterpart in the human being. It originates from the cranial end of the sternum deep to the pectoralis superficialis. The subclavius curves over the cranial edge of the scapula deep to the omotransversarius and trapezius to insert on the fascia that covers a deep shoulder muscle (the supraspinatus, see later) on the lateral side of the scapula as shown in Figs. 2-3 and 2-5.

D.2. Deep Muscles of the Shoulder

Reflect the trapezius and omotransversarius, and bisect and reflect the latissimus dorsi (Fig. 2-5). Find the subclavius and follow it to the craniolateral surface of the

24

scapula, where it inserts on a fascia that should be broken to expose the thick **supraspinatus**, which arises from the lateral surface of the scapula cranial to the scapular spine. An **infraspinatus** arises from the lateral surface of the scapula caudal to the scapular spine. You will have to bisect and reflect the deltoid to see it clearly. Both supraspinatus and infraspinatus muscles insert on the proximal end of

the humerus. The supraspinatus helps to protract the arm; the infraspinatus rotates it outward.

Trace the latissimus dorsi to its humeral insertion as it passes medial to the triceps (see Section E). The muscle fibers that arise from the caudal border of the scapula and insert with the latissimus dorsi constitute the **teres major**. This muscle helps to retract the arm. A

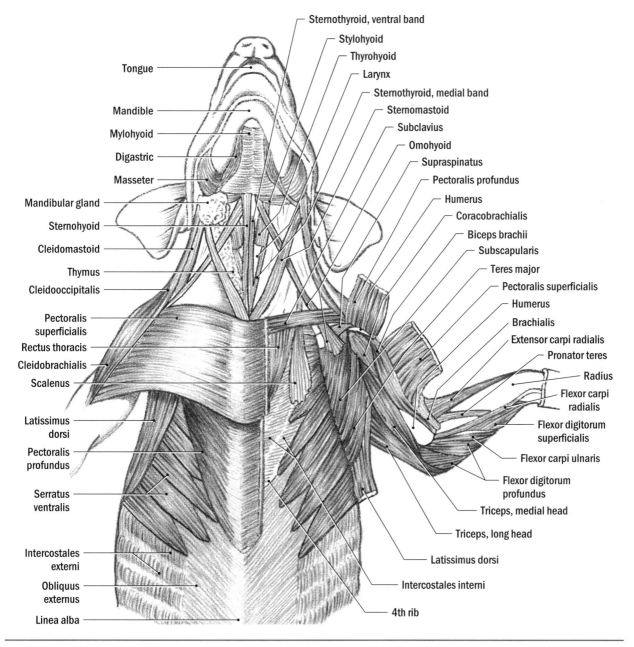

Muscles of the Shoulder

FIGURE 2-4
Ventral view of the muscles of the cervical, pectoral and axillary region of a fetal pig.
Deeper muscles are shown on the specimen's left side. The cutaneous muscles and the
tensor fasciae antebrachii, as well as the blood vessels and nerves of the brachial plexus, are not shown.

25

small **teres minor** arises from the border of the scapula between the infraspinatus and the lateral surface of the triceps. It is not easily seen and may be omitted. However, it can be found by separating the infraspinatus from the underlying triceps all the way to the caudoventral edge of the scapula. Lift the caudoventral border of the infraspinatus and follow it toward the shoulder joint. If you break the fascia that covers the ventromedial surface of the infraspinatus, you will discover a distinct muscle belly, the teres minor, that adheres to the infraspinatus. Proximal to the distal one-third of the scapula, the two muscles fuse.

The **rhomboideus** lies directly under the trapezius. It inserts on the dorsal, or vertebral, edge of the scapula on the medial side of the scapular cartilage, which may still be poorly developed in a fetus and look like a fold of connective tissue. The muscle can be subdivided into three poorly defined parts, depending on their origins. The **rhomboideus capitis** originates from the nuchal crest of the skull, the **rhomboideus cervicis** originates from the middorsal line together with the cervical part of the trapezius, and the **rhomboideus thoracis** arises from the spinous processes of the first six to eight thoracic vertebrae. All muscle parts act on the scapula, which they help to hold in place and to pull cranially or caudally. Human beings lack the rhomboideus capitis.

Cut through the muscular portions of the rhomboideus and pull the dorsal border of the scapula laterally. The **serratus ventralis** is the large, fan-shaped muscle arising by a series of slips from the ribs and from the trans-

*Muscles of
the Shoulder*

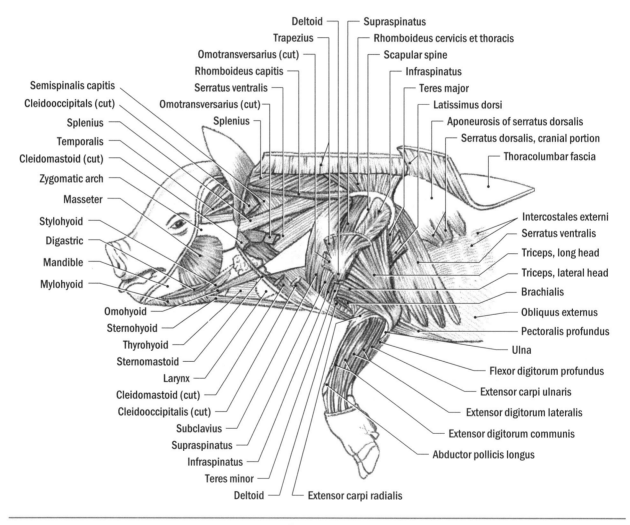

FIGURE 2-5
Lateral view of the deeper muscles of the neck and shoulder, and of the extensor muscles of the wrist and toes of a fetal pig.

verse processes of the cervical vertebrae. It attaches on the dorsal border of the scapula deep to the insertion of the rhomboideus. In addition to moving the scapula backward and forward, the serratus ventralis acts as a muscular sling that connects the trunk to the pectoral girdle and appendage. Recall that the pectoral girdle does not articulate directly with the vertebral column.

Turn the specimen on its back and bisect the pectoralis profundus, but not the subclavius (Fig. 2-4). Clean up the medial side of the scapula but do not injure blood vessels and a group of whitish strands (i.e., nerves of the brachial plexus) going to the arm. A neck muscle, the **omohyoid** (see Section F), attaches onto the fascia on the medial side of the scapula. Be careful not to injure it. Find again the serratus ventralis, latissimus dorsi, teres major, and the border of the supraspinatus in this view. The thick muscle arising from the medial surface of the scapula between the supraspinatus and teres major is the **subscapularis**. It inserts on the proximal end of the humerus and acts primarily as an adductor of the arm. Its caudal border is closely tied to the teres major and may be found only after tearing the fascia that covers the surface of the subscapularis.

E. MUSCLES OF THE ARM

The following muscles cover the humerus. Most arise from the humerus or scapula and move the forearm. Largest is the **triceps**, which you have already seen. The medial surface of the triceps is covered by the very thin **tensor fasciae antebrachii** (Fig. 2-3), a muscle band that originates from the caudodorsal corner of the scapula and by an aponeurosis along the ventral border of the latissimus dorsi in the axillary region. It inserts medially on the **antebrachial fascia**, which covers the medial surface of the arm, and, together with the triceps, on the olecranon of the ulna. Bisect it if it has not already been destroyed.

The triceps consists of three distinct heads—a **long head** from the caudal border of the scapula (Fig. 2-5), a **lateral head** from the proximal part of the lateral surface of the humerus, and a **medial head** from the proximal half of the medial surface of the humerus (Fig. 2-4). All converge to form a large, common tendon that passes over the elbow to insert on the proximal end of the olecranon of the ulna. The triceps is the extensor of the forearm.

Two smaller muscles lie on the cranioventral surface of the humerus, laterally a **brachialis** and medially a **biceps brachii** (Figs. 2-4 and 2-5). The insertions of the brachiocephalicus and pectoralis superficialis pass between them. The brachialis arises from the humerus and inserts on the ulna. The biceps brachii, which is two-headed in human

beings, arises in the pig by a single tendon that passes through a groove on the craniomedial surface of the humerus to attach to a small process (**coracoid process**) on the scapula. This process lies just medial and cranial to the glenoid cavity. The biceps brachii inserts on the radius and ulna. Both brachialis and biceps brachii are flexors of the forearm.

A small **coracobrachialis** (Fig. 2-4) arises from the coracoid process, crosses the medial surface of the shoulder joint, and inserts on the humeral shaft. It is a weak adductor of the arm, and it also reinforces the shoulder joint.

If you want to see the forearm muscles, you need to remove the tough and thick connective tissue enveloping the antebrachium (forearm) and foot. They can be divided into flexor muscles, which flex the wrist and toes, and extensor muscles, which extend the wrist and toes. On the caudomedial surface of the antebrachium and wrist (Fig. 2-4), you will find a pair of wrist flexors, namely, the medial **flexor carpi radialis** and the **flexor carpi ulnaris**, both of which insert on the carpals. You will also find a pair of toe flexors, the **flexor digitorum superficialis** and the **flexor digitorum profundus**, both of which insert on the phalanges with long tendons. On the craniolateral surface of the antebrachium and wrist (Fig. 2-5), you will find a pair of wrist extensors, the caudal **extensor carpi ulnaris** and the prominent cranial **extensor carpi radialis**. You will also find toe extensors, namely, an **extensor digitorum lateralis** with long tendons to the fourth and fifth toes, and an **extensor digitorum communis** with long tendons to all four toes. An **abductor pollicis longus** is found just proximal to the wrist, deep to the extensor carpi radialis and the extensor digitorum communis.

Lateral Neck Muscles

F. LATERAL NECK MUSCLES

Muscles on the lateral surface of the neck extend from the pectoral girdle and sternum to the skull. Their primary function is to flex and turn the head.

Again find the three muscles of the brachiocephalicus, which have been described with the shoulder muscles (see Section D.1). A narrower **sternomastoid** lies cranioventrally to the cleidomastoid and extends between the front of the sternum and the mastoid process of the skull.

Carefully dissect deep between the sternomastoid and the midventral region of the neck (Figs. 2-4 and 2-5). A large thymus lies on either side of the rather firm voice box, or larynx. The thymus is crossed by a very thin, band-shaped **omohyoid**, which extends between the hyoid bone at the base of the tongue and the fascia covering the medial side of the scapula. The omohyoid passes deep to the sternomastoid, brachiocephalicus, subclavius, and the

27

prominent external jugular vein. You have already seen its tendinous scapular attachment (see Section D). The sternomastoid, cleidomastoid, and omohyoid have a branchiomeric origin, in common with the trapezius.

G. THROAT MUSCLES

Several ribbon-shaped muscles extend from the sternum to the hyoid and larynx on the ventral side of the neck (Fig. 2-4). They belong to a subgroup of somatic muscles called the **hypobranchial group** because they have evolved from a group of muscles that is situated ventral to the gills in ancestral fishes. The hypobranchial muscles described here pull the larynx and hyoid bone caudally, an action that occurs in swallowing. The names of the muscles are descriptive of their origins and insertions. Most superficial of the group is a **sternohyoid**. If you have difficulty finding it, it may still be stuck to the internal surface of the platysma (see Section A). After you have identified its attachments and isolated it, push it to the side to reveal the underlying muscles and the trachea, or windpipe. A long **sternothyroid** lies deep to the sternohyoid and thymus and covers the thyroid gland and trachea. Midway between its origin and insertion, it splits into a ventral and a dorsal muscle band, which insert midventrally and laterally on the thyroid cartilage, one of the cartilages of the larynx. A shorter, but wider **thyrohyoid** extends from the insertions of the sternothyroid to the hyoid.

More cranial hypobranchial muscles, which will not be described, lie deep between the hyoid and chin and extend into the tongue. They pull the hyoid cranially, move the tongue, and help to open the jaw.

Trunk Muscles

H. HEAD MUSCLES

Head muscles are branchiomeric in origin. Most are responsible for jaw movements, and a few are involved in swallowing.

Remove the skin from the side of the skull between the eyelids and the auricle. Also remove a facial muscle extending from the skull to the anterior part of the auricle. A **temporal** muscle lies beneath it (Fig. 2-3). It arises from the temporal fossa of the skull and passes deep to the zygomatic arch to insert on the coronoid process of the mandible. A large **masseter** arises from the ventral margin of the zygomatic arch and inserts on the lateral surface and angular process of the mandible. The temporal and masseter are major jaw-closing muscles. Smaller **pterygoid** muscles arise deep from the base of the skull and

28

attach to the medial side of the jaw. They will not be seen in this dissection.

A **digastric** (Figs. 2-3 and 2-4) inserts along the medioventral margin of the mandible. Reflect but do not remove the mandibular salivary gland located at the angle of the mandible and trace the digastric cranially and dorsally. It forms a strong tendon that arises from the base of the skull. The digastric is the major jaw-opening muscle. It derives its name from its condition in human beings, in which the muscle consists of anterior and posterior bellies that are separated from each other by a short tendon.

The **mylohyoid** consists of a thin sheet of transverse muscle fibers forming part of the floor of the mouth between the digastric muscles of opposite sides of the body. It arises from the mandible deep to the insertion of the digastric and inserts on a midventral septum of connective tissue and on the hyoid. The mylohyoid raises the floor of the mouth and assists in swallowing.

I. TRUNK MUSCLES

Trunk muscles are somatic. They can be divided into an **epaxial group**, which lies dorsal and lateral to the vertebral column and acts primarily to brace the back and to move it from side to side as well as to straighten it, and a **hypaxial group**, which contributes to the musculature of the rib cage and forms the abdominal wall.

Three thin sheets of abdominal muscles arise dorsally from the ribs and thoracolumbar fascia and insert ventrally by aponeuroses to a midventral connective tissue septum, the **linea alba**. All three muscle layers help to support the viscera and to compress the abdominal cavity during expiration, coughing, and defecation. They can be distinguished by their position in relation to one another and by their distinctive muscle fiber orientation (Figs. 2-3, 2-4, and 2-6).

The superficial layer, the **obliquus externus**, extends cranially to the fourth rib; its muscle fibers run from craniolateral to caudomedial. Lift the caudal border of the pectoralis profundus and find the cranial border of the obliquus externus near the fourth rib. Now lift this cranial border and carefully separate the obliquus externus from the underlying connective tissue, muscle layer, and rib cage. With a pair of scissors, bisect the obliquus externus as far as it has been separated from the underlying tissue and reflect its halves. Continue to separate the obliquus externus from the underlying tissue in this way as far caudally as you can. Reflect the ventral half all the way to the linea alba and the dorsal half as far as you can. Notice that the obliquus externus becomes aponeurotic toward the pelvic and inguinal regions.

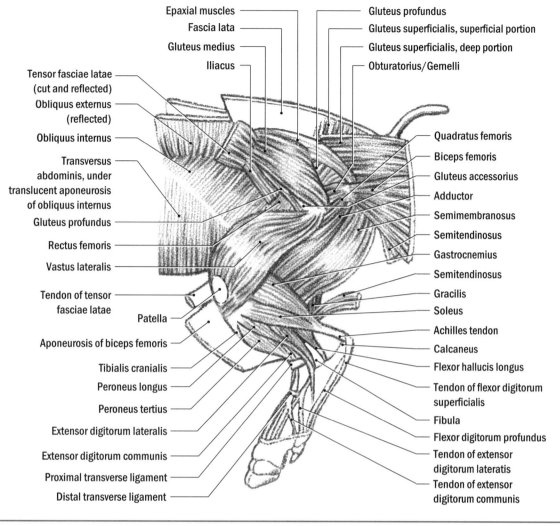

Epaxial muscles
Fascia lata
Gluteus medius
Iliacus

Gluteus profundus
Gluteus superficialis, superficial portion
Gluteus superficialis, deep portion
Obturatorius/Gemelli

Tensor fasciae latae
(cut and reflected)
Obliquus externus
(reflected)
Obliquus internus
Transversus
abdominis, under
translucent aponeurosis
of obliquus internus
Gluteus profundus
Rectus femoris
Vastus lateralis
Tendon of tensor
fasciae latae
Patella
Aponeurosis of biceps femoris
Tibialis cranialis
Peroneus longus
Peroneus tertius
Extensor digitorum lateralis
Extensor digitorum communis
Proximal transverse ligament
Distal transverse ligament

Quadratus femoris
Biceps femoris
Gluteus accessorius
Adductor
Semimembranosus
Semitendinosus
Gastrocnemius
Semitendinosus
Gracilis
Soleus
Achilles tendon
Calcaneus
Flexor hallucis longus
Tendon of flexor digitorum
superficialis
Fibula
Flexor digitorum profundus
Tendon of extensor
digitorum lateratis
Tendon of extensor
digitorum communis

FIGURE 2-6
Lateral view of the deeper pelvic and leg muscles of a fetal pig.

Trunk Muscles

The **obliquus internus** lies directly under the obliquus externus. It is muscular only in the dorsal half of the abdominal wall, and its muscle fibers run from caudodorsal to cranioventral, that is, crosswise to the muscle fiber direction of the obliquus externus (Fig. 2-6). The obliquus internus inserts by an aponeurosis to the linea alba. Because this aponeurosis is practically transparent, it is possible to identify the underlying **transversus abdominis** without any further dissection. Its muscle fibers run essentially dorsoventrally.

The **rectus abdominis** is a longitudinal muscle band that arises from the fourth rib cartilage and the sternum deep to the obliquus externus and attaches caudally to the pubis. To see it well, reflect the ventral half of the bisected obliquus externus and notice on its inner surface the rec-

tus abdominis as a band with longitudinal muscle fibers running along the midventral linea alba. It is active during ventral bending of the trunk.

Thoracic hypaxial muscles assist in respiration. A short **rectus thoracis** (Fig. 2-4) extends from the sternum across the cranial end of the rectus abdominis to attach to cranial ribs, which it pulls caudally. A **scalenus** lies laterally and dorsally to the rectus thoracis. It arises from cervical vertebrae, crosses the cranial part of the insertion of the serratus ventralis, and attaches to the first four cranial ribs, which it pulls forward.

Pull the vertebral edge of the scapula laterally and examine the chest wall medial to the insertion of the serratus ventralis. The muscle fibers that extend between two ribs and that run from craniodorsal to caudoventral **29** constitute the **intercostalis externus**. In the pig, but not inhuman beings, the intercostales externi are weak muscles and may be completely absent.

Carefully cut through one intercostalis externus, and you will find an **intercostalis internus**, with muscle fibers running at right angles to those of the intercostalis externus. You can also see the intercostalis internus ventral to the attachment of the scalenus on the rib cage (see Fig. 2-4), because the intercostales externi do not reach ventrally beyond the insertions of the serratus ventralis.

The thoracic layers correspond to the oblique layers in the abdominal wall. There is only a trace of the transverse layer (**transversus thoracis**) in the thoracic wall. It lies on the internal side of the thorax near the sternum and will be seen when the thorax is opened (Fig. 3-7).

Reflect the rhomboideus and latissimus dorsi, including the thoracolumbar fascia attaching to it (Fig. 2-5). You will find several muscular slips inserting on the dorsal part of ribs and originating by a long and wide aponeurosis from the middorsal line. This is the **serratus dorsalis**, which consists of two distinct parts. The cranial part consists of three to five slips attaching to the fourth to ninth ribs caudal to the serratus ventralis. Its muscle fibers run from dorsocranial to ventrocaudal. The caudal part of the serratus dorsalis may be separated from the cranial part by a space; its five to nine slips insert on the ninth to last ribs (Fig. 2-3). Its muscle fibers run from dorsocaudal to ventrocranial.

The action of the muscles of the thoracic wall is primarily to stabilize the ribs and maintain the integrity of the thoracic wall as the diaphragm(Exercise 3) contracts, pushes the abdominal viscera caudally, expands the thoracic cavity, and draws air into the lungs. This process is called inspiration. The expulsion of air (expiration) results primarily from the elastic recoil of the lungs, abdominal viscera, and thoracic wall.

The epaxial trunk muscles lie along the dorsal surface of the vertebral column. You may be able to expose them by bisecting and reflecting the serratus dorsalis and thoracolumbar fascia, but they are subdivided in a complex manner by connective tissue sheets and are difficult to dissect. Deep to the rhomboideus and serratus dorsalis, you will find the **splenius**, with its separate insertions to the nuchal crest, the zygomatic arch, and the wing of the atlas (Fig. 2-5). The splenius helps to raise the head or turn it to the side.

J. MUSCLES OF THE PELVIS AND THIGH

Unlike the muscles in the pectoral region, which are arranged so that all but a few muscles act across a single joint, many of the pelvic muscles extend across both the hip and knee joints and, therefore, can move the thigh and the shank at the same time. When the hind foot is

held on the ground, these muscles move the trunk relative to the foot.

J.1. Lateral Muscles of the Hip and Thigh

Six muscles can be seen on the lateral surface of the hip and thigh (Fig. 2-3). A **tensor fasciae latae** arises from the iliac crest and by an aponeurosis that is fused to the underlying muscle (the gluteus medius, see later) and fans out to insert on the transparent **fascia lata** near the knee. Because a part of the fascia lata, in turn, attaches to the tibia, the tensor fasciae latae helps to extend the shank as well as to protract the thigh at the hip joint.

A very broad **biceps femoris** covers most of the lateral surface of the thigh caudal to the fascia lata. It arises from the caudal part of the sacrum and ischium and inserts by an aponeurosis along most of the length of the tibia. It forms the lateral wall of the **popliteal fossa**, the depression behind the knee joint. The biceps femoris acts across both the hip and knee joints and, thus, retracts the thigh and flexes the shank.

The **gluteus superficialis** lies between the tensor fasciae latae and the biceps femoris. It is a complex muscle in the pig because it is partly fused to the biceps femoris, forming a gluteobiceps. If time permits, the following dissection can be attempted. Cut through the fascia lata along the dorsal border of the biceps femoris and separate the fascia carefully from the underlying muscles as far dorsally as you can. When you reach the tensor fasciae latae, make a longitudinal cut through the fascia lata along the dorsal border of the tensor fasciae latae. As you lift the fascia and hold it against a light, you will discover in it a very thin muscle sheet, which inserts on the fascia and arises from the sacrum. This muscle sheet constitutes the **superficial portion** of the gluteus superficialis (Figs. 2-3 and 2-6).

Now lift the dorsal border of the biceps femoris and separate its inner surface from the underlying muscles. You may need to cut through some blood vessels and smaller nerves but do not injure the very large nerve and blood vessels running toward the foot behind the knee joint. The origin of the **deep portion** of the gluteus superficialis is from the sacral and first two caudal vertebrae and is separated from the origin of the biceps femoris by the longitudinal bundle of epaxial muscles. The deep portion of the gluteus superficialis soon crosses over deep to the gluteus femoris and fuses with it (Figs. 2-3 and 2-6).

The gluteus superficialis is quite distinct in human beings and forms the large buttock muscle (the gluteus maximus). The **gluteus medius** is more conspicuous in the pig than in the human being (Fig. 2-6). It lies deep to the tensor fasciae latae, gluteus superficialis, and biceps femoris. It will be seen better later.

A band-shaped **semitendinosus** lies caudal to the biceps femoris (Fig. 2-3). It arises from the ischium and is fused to the biceps femoris near its origin. It inserts on the medial side of the tibia, where it helps to form the medial wall of the popliteal fossa and gives off a tendon to the **Achilles tendon**. You can feel its powerful tendon of insertion on the caudomedial border of the popliteal fossa in your own leg. The semitendinosus assists the biceps femoris in thigh retraction and shank flexion. A semimembranosus lies caudal and medial to the semitendinosus, but it can be seen more clearly in a medial dissection of the thigh (see Section J.2 and Figs. 2-6 and 2-7).

Bisect and reflect the biceps femoris and the tensor fasciae latae, if you have not already reflected the latter muscle during the dissection of the gluteus superficialis (Fig. 2-6). The gluteus medius can now be seen clearly. It arises from the lateral surface of the ilium and sacrum. Its muscle fibers pass caudoventrally to insert on a large bump (the greater trochanter) on the proximal end of the femur caudodorsal to the hip joint. This insertion enables the gluteus medius to retract the thigh, as well as to act as a thigh abductor. Carefully separate it from the underlying **gluteus accessorius**, which also arises from the ilium and sacrum but inserts on the greater trochanter and also on the femur itself, deep to the origin of the vastus lateralis (see later and Fig. 2-6). Now bisect and reflect the gluteus medius to expose the entire gluteus accessorius. The **piriformis**, a distinct muscle in many mammals and pear-shaped in human beings, is completely fused with the gluteus medius in the pig.

Bisect and reflect the gluteus accessorius. The large **ischiatic nerve**, which supplies many of the hip and leg muscles, lies deep to the gluteus accessorius and superficial to a large, fan-shaped muscle, the **gluteus profundus** (Fig. 2-6), which corresponds to the gluteus minimus in human beings. The gluteus profundus arises from the entire length of the ilium and from the ischium cranial and dorsal to the acetabulum and ischiatic nerve. Its muscle fibers converge to insert on the greater trochanter distal to the insertion of the gluteus accessorius. This muscle, too, is an adductor of the thigh.

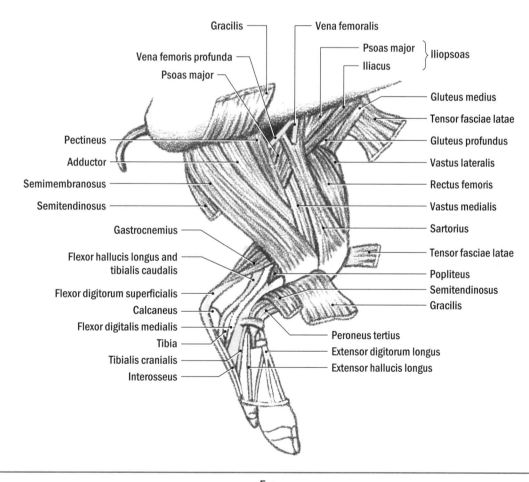

Muscles of the Pelvis and Thigh

FIGURE 2-7
Medial view of the hind leg muscles of a male fetal pig.

Caudal to the gluteus profundus and deep to the ischiatic nerve, you will find smaller hip muscles, which rotate the thigh and assist in its retraction. The **obturatorius** and **gemelli** muscles, which are distinct muscles in many mammals and in human beings, are fused in the pig (Fig. 2-6). They can be found caudal to the gluteus profundus. They arise from the region of the obturator foramen of the pelvis and from the ischial tuberosity, and insert in the fossa between the greater trochanter and the femur. The **quadratus femoris** originates from the ischium and inserts on the caudal surface of the femur.

J.2. Medial Thigh Muscles

A thin, broad, muscular sheet, the **gracilis**, covers the caudomedial half of the thigh (Fig. 2-7). The gracilis arises from the pubic symphysis and from the surface of a deeper thigh muscle (the adductor, see later) and inserts by an aponeurosis along most of the length of the tibia. It is an adductor and retractor of the thigh and a flexor of the shank.

The **sartorius** is a superficial, narrow, and thin band of muscle cranial to the gracilis and caudal to the medial border of the tensor fasciae latae. If you have not done so before, separate the tensor fasciae latae from the sartorius and reflect it (Fig. 2-7). The sartorius arises by two heads, which are separated by femoral blood vessels, from the ilium and pubis and from adjacent muscles. Distally, it unites with the gracilis and inserts on the proximal end of the tibia. It helps to adduct the thigh and extend the shank.

Muscle Tissue

Bisect and reflect both the gracilis and sartorius. Most of the caudomedial part of the thigh is occupied by the large **semimembranosus** (Fig. 2-7), part of which has been seen previously from the lateral surface. The semimembranosus arises from the caudal part of the ischium and inserts on the distal end of the femur and on adjacent parts of the tibia. Together with the semitendinosus, it forms the medial wall of the popliteal fossa. It acts primarily as a retractor of the thigh.

A triangular **adductor** lies cranial to the semimembranosus. It arises from the ventral surface of the pubis and ischium, and its muscle fibers converge to insert along much of the length of the femur. It is not subdivided in the pig as it is in human beings and many other mammals. As its name implies, it adducts the thigh, but it also assists in thigh retraction.

A smaller, triangular **pectineus** lies cranial to the adductor. It arises from the front of the pubis and inserts on the femur just cranial to the insertion of the adductor. It too adducts the thigh.

An **iliopsoas** can be seen cranial and deep to the pectineus and sartorius (Fig. 2-7). It can be subdivided into two muscles in the pig. The larger, more medial **psoas major** originates from the ventral surface of the lumbar vertebrae, which cannot be seen at this point. It emerges through the abdominal wall to insert deep to the pectineus

on the lesser trochanter (a small bump on the femur distal to the greater trochanter of the femur). The more lateral **iliacus** originates from the ventral border of the ilium medial to the tensor fasciae latae and inserts in common with the psoas major. These muscles help to protract and rotate the thigh.

J.3. Cranial Thigh Muscles

The cranial part of the thigh is occupied by a large muscle complex, the **quadriceps femoris**, which is the major extensor of the shank. The four components of this muscle complex converge to form a common **patellar tendon**, which crosses the front of the knee joint to insert on the proximal end of the tibia. This tendon slides easily across the knee joint because it contains a sesamoid bone, the knee cap, or **patella**.

One of the heads of the quadriceps femoris, the **vastus lateralis,** can be seen on the lateral surface of the thigh (Fig. 2-6) because it arises from the greater trochanter and from much of the lateral surface of the femur. In a medial view, the **vastus medialis** can be seen arising from the medial surface of the femur deep to the sartorius (Fig. 2-7).

The **rectus femoris** lies between these two vasti muscles. Because it takes its origin from the pelvis just cranial to the acetabulum, it passes over both the hip and knee joints and also helps to protract the thigh. The small fourth head, the **vastus intermedius**, arises from the craniolateral surface of the femur. It can be found by dissecting between the vastus lateralis and the rectus femoris.

Shank muscles are not described here, but many can relatively easily be seen if the substantial subcutaneous connective tissue is removed from the shank and tarsus. Like their counterparts in the foreleg, the muscles of the shank, tarsus, and hind toes can be subdivided into flexors and extensors of the tarsus and the digits. They are, however, more numerous, more complex, and usually larger (Figs. 2-6 and 2-7).

K. MUSCLE TISSUE

K.1. Skeletal Muscle

Examine a histological slide preparation of a longitudinal section through **skeletal muscle**. The individual muscle cells are cylindrical and very long, with a uniform diameter ranging from 10 to 100 micrometers. Because of their great length, these muscle cells are also called muscle fibers. In a few cases, the **muscle fibers** may be as long as the muscle they form, but usually they are only a few centimeters long and may be connected in series by connective tissue to extend over the entire length of a muscle. Because these muscle fibers develop embryonically by an end-to-end fusion of individual embryonic muscle cell precursors (**myoblasts**), which each have one nucleus, each muscle fiber contains many nuclei. The nuclei lie close to the inner surface of the cell membrane, which is called **sarcolemma** in muscle fibers (Fig. 2-8A, B).

The cytoplasm of muscle fibers, called **sarcoplasm**, consists almost solely of **myofibrils**, which are the contractile elements of the muscle. They are barely visible with light microscopy because they are only 1 or 2 micrometers in diameter (Fig. 2-8A, B). The alternating dark and light bands that occur along a myofibril are known as **A bands** and **I bands**, respectively, because of their anisotropic and isotropic properties in polarized light. The A and I bands of the various myofibrils in a particular muscle fiber are in register with one another, so the entire muscle has a banded, or striated, appearance. Because of this appearance, skeletal muscles are often called **striated muscles**.

Electron microscopic and biochemical studies have shown that the myofibrils, in turn, are composed of many ultramicroscopic myofilaments of two types and that

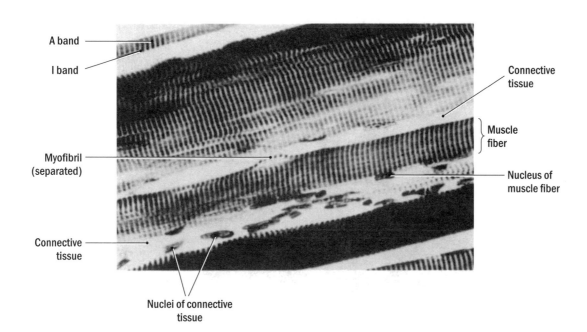

Muscle Tissue

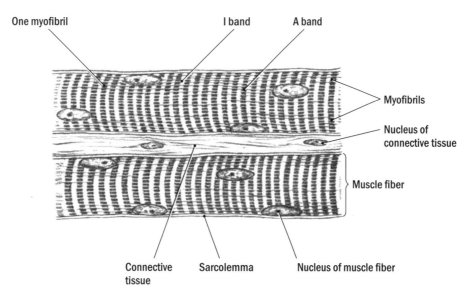

FIGURE 2-8
Skeletal muscle.
(A) Photomicrograph at high magnification of a longitudinal section through skeletal muscle fibers.
(B) Diagrammatic enlargement of a portion of two skeletal muscle fibers.

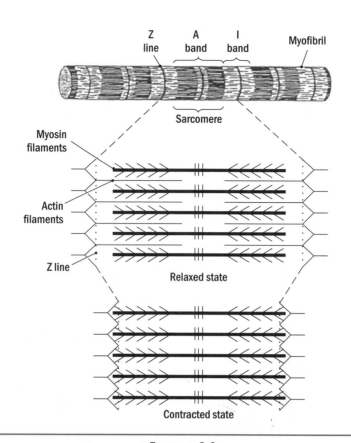

FIGURE 2-8 CONTINUED
Skeletal muscle.
(C) Diagrammatic view of a myofibril at the microscopic and electron microscopic level.
(C is redrawn from Walker, W. F., Jr., and Homberger, D. G. *Vertebrate Dissection*, 8th ed.
Philadelphia: Saunders College Publishing, 1992. After Dorit, R. L., Walker, W. F., Jr., and Barnes,
R. D. *Zoology*. Philadelphia: Saunders College Publishing, 1991. And Fawcett, D. W. *Bloom and
Fawcett, A Textbook of Histology*, 12th ed. New York: Chapman & Hall, 1994.)

Muscle Tissue

these are subdivided into contractile units called **sarcomeres** (Fig. 2-8C). The thicker **myosin filaments** are limited to the dark A bands within a sarcomere. The thinner **actin filaments** occupy the light I bands. Each I band is divided by a **Z line** which serves as an anchoring place for the actin filaments of two adjacent sarcomeres. The actin filaments extend into the A bands where they interdigitate with the myosin filaments. When a muscle fiber contracts, complex biochemical interactions between the two types of myofilaments cause the actin filaments to slide deeper into the array of myosin filaments. Consequently, the I bands become narrower, and the myofibrils—and with them the muscle fiber—generate tension and tend to shorten. The precise geometry of the myofilaments is responsible for the great speed and force with which striated muscles contract.

Skeletal muscles are innervated by myelinated nerves that travel within the connective tissue permeating them. Each nerve fiber divides into several terminal branches, each of which attaches to the sarcolemma of one muscle fiber through a **motor end plate** (or myoneural junction). All the muscle fibers that are innervated by a single nerve fiber form a **motor unit.** When a skeletal muscle contracts, only about one-third of all motor units contract at a given time.

K.2. Cardiac Muscle
Examine a histological slide preparation of a longitudinal section through the **cardiac muscle** of the heart wall (Fig. 2-9A). This muscle type also is striated, but its striations are much less obvious than those in skeletal muscle. The nuclei are situated in the center of the cells. The individual muscle cells have a diameter of about 15 micrometers. With a length of about 85 to 100 micrometers, they are not as long as the muscle fibers of skeletal muscles. The muscle cells

branch and are joined to other muscle cells at specialized junctions called **intercalated disks** (Fig. 2-9B).

The individual muscle cells are not innervated by nerves. Action potentials initiated by the pacemaker of the heart (see Exercise 4) spread rapidly through the muscle cells themselves, crossing from cell to cell at the intercalated disks. As a consequence, the cardiac muscle tends to contract as a unit.

K.3. Smooth Muscle

Examine a histological slide preparation of a section through **smooth muscle** (Fig. 2-10A). Smooth muscle is found in the walls of most visceral organs and blood vessels. Smooth muscle fibers are elongated, spindle-shaped cells with a length from 20 to 500 micrometers and a single nucleus in their center (Fig. 2-10B). The muscle fibers usually are tightly packed and staggered, so the

Muscle Tissue

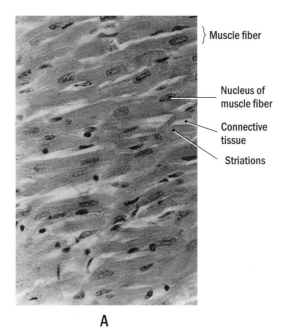

} Muscle fiber

Nucleus of muscle fiber

Connective tissue

Striations

A

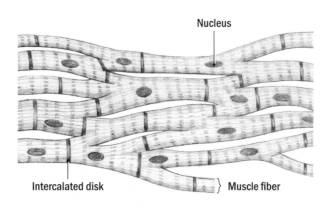

Nucleus

Intercalated disk

} Muscle fiber

B

FIGURE 2-9
Cardiac muscle.
(A) Photomicrograph at high magnification of a longitudinal section through cardiac muscle, in which the intercalated disks are not visible.
(B) Diagrammatic enlargement of cardiac muscle fibers with intercalated disks.

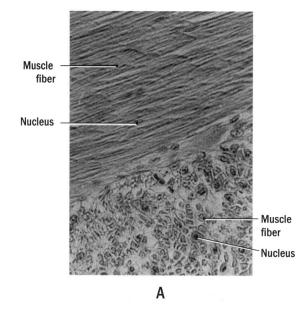

Muscle
fiber

Nucleus

Muscle
fiber

Nucleus

A

Muscle Tissue

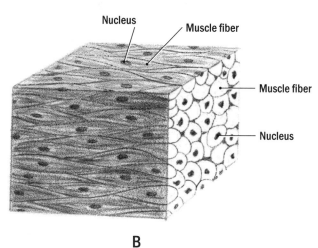

Nucleus

Muscle fiber

Muscle fiber

Nucleus

B

FIGURE 2-10
Smooth muscle. (A) Photomicrograph at high magnification of a longitudinal (upper half)
and transverse section (lower half) through smooth muscle. (B) Diagrammatic enlargement of smooth muscle fibers.
(B is redrawn from Junqueira, L. C., Carneiro, J., and Kelley, R. O. *Basic Histology*, 7th ed. Norwalk, CT: Appleton and Lange, 1992.)

tapered ends of some muscle fibers overlap the thicker center of other muscle fibers. As a consequence, the diameter of smooth muscle fibers varies considerably when the tissue is viewed in cross section. Each muscle fiber contains myofibrils, but these are barely visible by light microscopy. The light and dark bands of the myofibrils within a particular muscle cell are not in register with one another, therefore, the muscle cells do not appear striated.

Nerve fibers terminate on only some of the muscle fibers, so any action potential spreads from the innervated muscle cells to the rest of the muscle cells. However, it does so more slowly than in cardiac muscle. Smooth muscle is well suited for the slow but sustained contractions needed to move food through the digestive tract and to control peripheral blood flow and pressure.

Digestive and Respiratory Systems

O RGANISMS ACQUIRE RAW MATERIALS and release waste products through the digestive and respiratory systems. Specifically, the respiratory system acquires oxygen and releases carbon dioxide (CO_2), a waste product from cell respiration. The digestive system acquires nutrients, such as foods, vitamins, and minerals, and releases feces consisting of indigestible materials, such as roughage and bacteria. Bile pigments, a product from the breakdown of senescent red blood cells and the hemoglobin they contain, are also voided with the feces. The nitrogenous wastes of cellular metabolism are removed by the excretory system, which will be studied later (see Exercise 5).

The digestive and respiratory systems are usually studied together because their parts are connected and found adjacent to one another in terrestrial vertebrates. This anatomical proximity is the result of the embryonic development of the lungs as an outgrowth from the pharyngeal part of the digestive tract.

A. HEAD AND NECK

A.1. Salivary Glands

If you did not see the salivary glands when dissecting the muscles, carefully remove the skin from one side of the head and neck over the area shown in Fig. 3-1 to expose them, or study the glands in a demonstration dissection. If you do the dissection yourself, first remove the skin while leaving the underlying facial musculature in place. You must do this dissection very carefully because the facial muscles attach to the skin (see Exercise 2). Muscle tissue can be recognized by the small, parallel bundles of muscle fibers that can be seen when the connective tissue is carefully picked off. Then remove only the facial muscle layer, carefully, to avoid injuring the underlying salivary glands. Glandular tissue has a texture that is different from muscle tissue and consists of little lobules of tissue clustered in bunches.

The **parotid gland** is a large, thin, approximately triangular gland occupying the area of the upper neck between the base of the ear, the shoulder, and the angle of the lower jaw. The angle of the lower jaw is covered by a large jaw muscle, the **masseter**. The **parotid duct** leads rostrally from the rostroventral corner of the parotid gland along the ventral edge of the masseter and opens into the mouth cavity on a papilla on the inner, or buccal, surface of the upper lip dorsal to the angle of the mouth. Do not confuse the parotid duct with nerves.

The branches of the **facial nerve** emerge from beneath the rostral border of the parotid gland; in contrast, the parotid duct arises from the very edge of the parotid gland. Furthermore, nerves are yellowish white, solid, and far tougher than the hollow and flattened parotid duct. The facial nerve supplies the facial muscles. One branch of the facial nerve follows the parotid duct; another crosses the masseter more dorsally.

Around the rostroventral corner of the parotid gland, you may find several distinct bead-like globules of tissue that have a smooth surface and cling to the larger blood vessels. These globules are **lymph nodes**. Most of the **mandibular gland** lies beneath the parotid gland and just caudal to the angle of the jaw. Its body is sausage shaped, and it has a slightly darker color and larger lobules than the parotid gland. Its rostral end lies medial to the mandible.

The **mandibular duct** passes medially to the angle of the jaw and can be followed rostrally by carefully bisecting some small jaw muscles (the digastric and mylohyoid, see Exercise 2, Section H) under which it passes. Do not injure the prominent **lingual nerve**, which crosses the mandibular duct from caudolateral to rostromedial. The mandibular duct opens on a papilla on the sublingual floor of the mouth (see later; Section A.2).

Thin, elongated, yellowish, and finely granulated strips of glandular tissue cling to the mandibular duct and constitute the **sublingual glands**. Although there are two sublingual glands, it is difficult to identify them individu-

ally. The monostomatic, or "one-orificed," sublingual gland ends in a duct that follows the mandibular duct and opens together with it on the sublingual floor of the mouth (see Fig. 3-1). It is usually too small to see with the naked eye or even low magnification. The other, polystomatic, or "many-orificed," sublingual gland continues rostrally beyond the previous gland and opens with multiple very short ducts independently on the sublingual floor of the mouth.

The **dorsal** and **ventral buccal glands** can be seen as strips of granulated tissue on the surface of the **buccinator muscle**. This muscle forms the muscular portion of the cheek wall between the angle of the mouth and the masseter (see Fig. 3-1). The buccal glands open with numerous minute ducts on the inner, or buccal, surface of the upper and lower lips near the angle of the mouth.

The secretions of all these glands combine to produce saliva, a complex solution containing a mixture of amylase (ptyalin), an enzyme that initiates the breakdown of complex carbohydrates (such as starch); chloride ions, which are necessary for the action of ptyalin; water; mucus, which lubricates the food and facilitates the swallowing process; and substances that help control the bacterial flora in the mouth cavity. As in sweat glands, myoepithelial cells assist in the discharge of the secretions from the salivary glands.

A.2. Mouth, Pharynx, Larynx, and Neck

Expose the organs of the mouth and pharynx by inserting a pair of scissors in the angle of the mouth on the side not previously dissected and cutting caudally. If your specimen is large, you may need to cut through the cheek on both sides. Spread open the mouth and, as you cut through the cheek, make the incision follow the curvature of the tongue. Do not cut into the roof of the mouth cavity. Continue the incision until you see a little flap of tissue, the epiglottis, projecting dorsally behind the free border of the soft palate. Carefully pull the epiglottis down and continue your incision dorsal to it and on into the gullet, or esophagus. You have cut far enough when you can see the entire larynx with its opening, the glottis (see Fig. 3-2). The mouth cavity and pharynx can now be swung open.

Certain **teeth** may have emerged through the gums; others may still remain beneath the gums. You are likely to see the third incisor and the canine teeth in the upper and lower jaws. You can palpate the gums to locate the other teeth that have not yet broken through. In adult mammals, teeth play a crucial role in the acquisition and mechanical breakdown of the food. But in baby animals, teeth may hurt the mother's nipples and induce her not to let her young suckle.

The **tongue** of mammals is completely muscular and does not contain any bony or cartilaginous elements. The lingual muscles attach caudally to the hyoid bone (see later) and rostrally to the angle of the mandible (see Fig. 3-3). The tongue manipulates the food, places it between

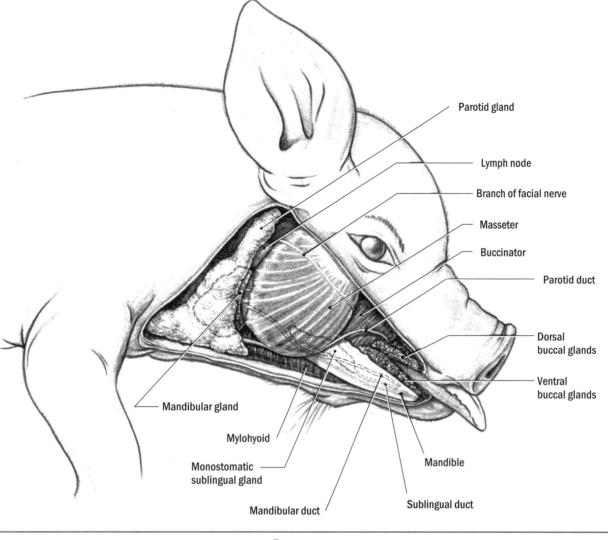

Parotid gland

Lymph node

Branch of facial nerve

Masseter

Buccinator

Parotid duct

Dorsal
buccal glands

Ventral
buccal glands

Mandibular gland

Mylohyoid

Monostomatic
sublingual gland

Mandibular duct

Mandible

Sublingual duct

Head and Neck

FIGURE 3-1
Lateral view of the salivary glands, their ducts, and the branches of the facial nerve of a fetal pig.
The monostomatic sublingual gland, mandibular duct, and sublingual duct lie behind, i.e., medial to the mandible.

the upper and lower tooth rows of the cheeks for chewing, mixes it with saliva, shapes it into a bolus, and finally pushes it into the gullet for swallowing.

In piglets, the tongue also has a crucial task during suckling; it establishes an airtight seal between the nipple of the sow and the mouth cavity of the piglet. The margins of the rostral one-third of the tongue bear rows of lacelike **marginal papillae**. These papillae apparently become engorged with blood and stiffen during suckling, presumably to create a firm connection between the tongue of the piglet and the nipple of the sow. They are lost at weaning.

The rostral half of the tongue surface is strewn with individually distinguishable, but tiny, buttonlike **fungiform papillae**. The caudal quarter of the tongue surface is densely covered with caudally pointing **filiform papillae**. A pair of flat **vallate papillae** are situated just rostrally to the filiform papillae. Microscopic **taste buds** are associated with the fungiform and vallate papillae, but the filiform

papillae only serve to provide a better grip on a food bolus as it is pushed into the gullet during swallowing.

Lift the tip of the tongue and look at the **sublingual floor of the mouth**. Where the sublingual floor of the mouth meets the underside of the tongue, you will discover a pair of tiny epithelial flaps, the **sublingual caruncles**, under which the sublingual and mandibular ducts open. Extending caudally from the sublingual caruncles on each side of the tongue are the **sublingual folds** onto which the many ducts of the polystomatic sublingual gland open. All these glandular orifices are too small to be visible with the naked eye or low magnification.

The palate consists of two parts. The **hard palate** with its transverse ridges, or **rugae**, is supported by the bony palate of the skull and forms the roof of the mouth, or **oral cavity**. The **soft palate** extends caudally from the hard palate. It has a smooth surface and is completely muscular, without any skeletal support. It forms the roof **39**

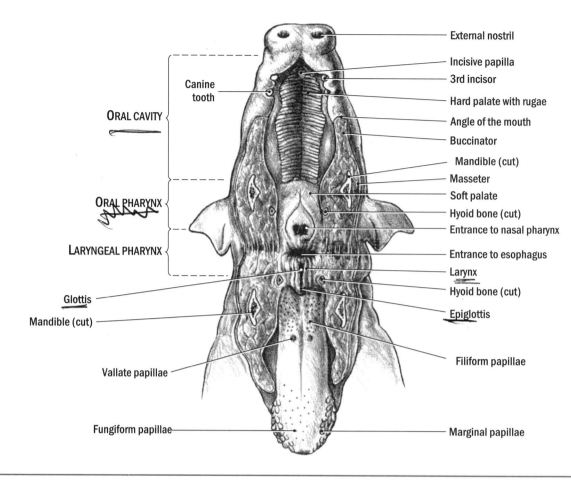

Canine tooth

ORAL CAVITY

ORAL PHARYNX

LARYNGEAL PHARYNX

Glottis

Mandible (cut)

Vallate papillae

Fungiform papillae

External nostril
Incisive papilla
3rd incisor
Hard palate with rugae
Angle of the mouth
Buccinator
Mandible (cut)
Masseter
Soft palate
Hyoid bone (cut)
Entrance to nasal pharynx
Entrance to esophagus
Larynx
Hyoid bone (cut)
Epiglottis
Filiform papillae
Marginal papillae

Head and Neck

FIGURE 3-2
The oral cavity and pharynx of a fetal pig. Cuts have been made through the cheek and mandible on both sides, and the tongue and floor of the mouth have been pulled away from the palate and roof of the mouth. (After a drawing by P. Anne Smith.)

of the oral pharynx (see later). A prominent **incisive papilla** can be seen at the rostral end of the hard palate. You may be able to discern the minute orifices of the **incisive ducts** on each side of the incisive papilla. These incisive ducts pass through fissures in the bony palate of the skull and lead to the paired **vomeronasal organs**. These are minute canals that end blindly. They are embedded in the floor of the nasal cavities (see later) and serve as accessory olfactory organs, which pick up pheromones that serve for intraspecific, interindividual communication in almost all mammals. The vomeronasal organs are relatively small in pigs and human beings and are absent in the aquatic whales and porpoises.

Paired **nasal cavities** lie dorsal to the hard palate and can be seen in a demonstration of a sagittal section of the head (see Fig. 3-3). A vertical **nasal septum** separates the two nasal cavities, and each cavity is largely filled with folds of tissue, the **conchae**, which increase the surface area available for olfaction and for humidifying and warming the inspired air. Further information on the nose is presented in Exercise 6.

Both the nasal and oral cavities lead caudally into the **pharynx**, which is subdivided into three parts. The **oral pharynx** is the caudal extension of the oral cavity and lies ventral to the soft palate. The **nasal pharynx** is the caudal extension of the nasal cavities and lies dorsal to the soft palate. Both the nasal pharynx and the oral pharynx lead into the **laryngeal pharynx** caudal to the soft palate and above the voice box, or larynx (see later). The pharynx in mammals has evolved from that part of the digestive tract of an ancestral fish into which the gill slits opened. In the embryos of mammals, certain branchial, or visceral, pouches develop; however, these pouches regress or give rise to other structures as the embryo develops.

Insert the blunt end of a pair of scissors into the exit of the nasal pharynx and make a longitudinal incision through the soft palate all the way to the hard palate. The nasal pharynx is a single chamber; it is formed by the confluence of the paired cranial nasal cavities. The **internal nares** (choanae) are the openings between the nasal cavities and the nasal pharynx; they are seen more clearly on a skull (see Exercise 1).

Near the middle of each lateral wall of the nasal pharynx is the slitlike opening of the **auditory tube**, which leads to the middle ear cavity. The middle ear cavity is filled with air. It is necessary to keep the air pressure in the middle ear cavity equal to that of the ambient air so that the delicate tympanic membrane is protected against tension caused by unequal pressures on it and is able to respond to sound waves. The auditory tube allows air to move between the middle ear cavity and the nasal cavity and, thereby, to adjust the air pressure within the middle ear cavity to that of the ambient air. The auditory tube and the middle ear cavity have developed embryonically from the first pharyngeal pouch, and this pharyngeal pouch has evolved from the first gill slit of ancestral fishes.

Return to the laryngeal pharynx, which leads both to the gullet, or esophagus, and to the windpipe, or trachea, via the larynx (see later for the specific description of these organs). It is in the laryngeal pharynx that the food and air passages cross. Air enters the nares, passes through the nasal cavities, continues through the nasal and laryngeal pharynges, and enters the trachea through the larynx. Food and water enter the mouth opening, pass through the oral cavity, continue through the oral and laryngeal

pharynges, and enter the esophagus, which lies dorsal to the trachea (see later and Fig. 3-3).

To see this area and its organs more clearly, first make a midventral cut (incision 1, Fig. 3-4) from near the chin to the front of the sternum. An incision somewhat like this one may already have been made to inject the blood vessels. If so, simply extend it. If you have already dissected the neck muscles of your specimen, you do not need to do any further dissection. Carefully extend the incision through the skin, if you have not removed it yet, and muscular tissue between the incision on the neck and the one you made to open the mouth cavity and pharynx.

As you do this dissection, you will expose the **thymus** (Fig. 3-7). It is a large glandular mass on each side of the neck, which develops from the embryonic tissue of certain branchial pouches. It is relatively large in a fetus and infant, but regresses with age and has usually completely disappeared by the time sexual maturity is reached. It has an important role in the development of the body's defense mechanisms, because certain types of white blood cells (T-lymphocytes) mature here before they are distributed to lymph nodes and other lymphoid tissues. T-lymphocytes are responsible for immune reactions associated with

Head and Neck

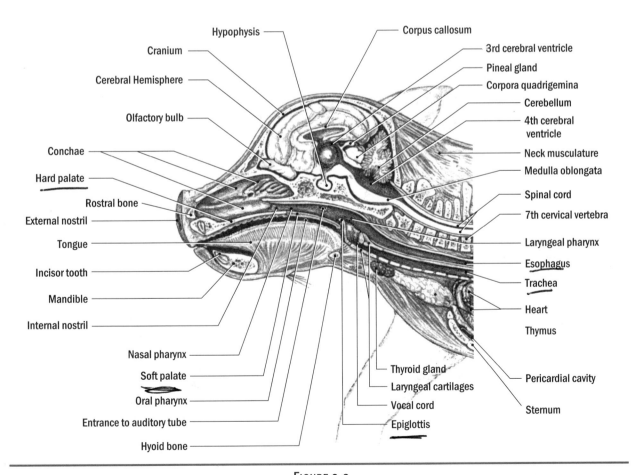

FIGURE 3-3
Sagittal section through the head and neck of a fetal pig.

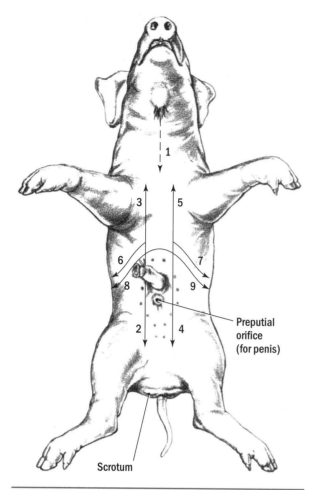

Head and Neck

Scrotum

Preputial
orifice
(for penis)

FIGURE 3-4
Ventral view of a male fetal pig, showing the incisions that
should be made to open the neck, chest, and abdomen.
Arrows indicate the direction in which these incisions
should be made, and numbers indicate the sequence
in which they should be performed.

fungal and viral infections, and for rejecting foreign tissue. The thymus also is necessary for the normal development of lymph nodes and other lymphatic organs, such as the spleen. It appears to exert its influence by the production of a hormonelike substance called lymphopoietin.

The **larynx** is supported by several large **laryngeal cartilages** that are part of the visceral skeleton. There is a trough-shaped flap of tissue, the **epiglottis**, at the rostral end of the larynx. The laryngeal skeleton is connected to the **hyoid bone** (see Exercise 1). You may be able to see or palpate part of the hyoid bone sticking out from the section you made to open the oral cavity (see Fig. 3-2). Air enters the larynx easily when its **glottis**, a longitudinal slitlike opening in the center of the larynx, is open during inspiration. The windpipe, or **trachea**, extends caudally from the larynx. It is held patent by oval cartilaginous rings in its walls, so air can easily move through it during respiration.

During swallowing, food is prevented from entering the larynx and trachea by closure of the glottis and a folding back of the epiglottis over the glottis. Food is pushed into the usually collapsed **esophagus**, which is the compressed tube dorsal to the trachea, by a backward movement of the tongue and a rhythmic contraction of the muscular walls of the pharynx. You can follow this swallowing process by putting your fingers on your own larynx, or Adam's apple, while swallowing.

Insert a pair of fine scissors into the glottis and cut caudally through its dorsal wall and into the trachea. Spread the two halves of the larynx slightly apart and observe the pair of small folds of whitish tissue on the inside of the lateral walls of the larynx. These folds, which extend dorsally from the base of the epiglottis, are the **vocal cords**. They are much better developed in an adult and may not be visible yet in a fetal pig. Laryngeal muscles move the vocal cords closer together and apart and control their tension. The mechanism of vocalization is not based on the vibration of the vocal cords, like the vibration of violin strings, as was believed earlier. The vocal cords actually create sound by opening and closing the glottis and thereby creating puffs of air during exhalation, in a manner similar to that by which clapping your hands creates sound.

Trace the trachea caudally, being careful not to injure the larger veins and arteries that lie beside it and at the base of the neck. A short distance caudal to the larynx, the trachea is covered by a purplish, compact, glandular mass, the **thyroid gland** (Fig. 3-3). The thyroid gland is one of many **endocrine glands** whose secretions, known as hormones, are discharged directly into the circulatory system rather than indirectly through ducts to the surface of the body or into the cavity of an organ, as is the case for **exocrine glands** such as salivary glands. The thyroid gland produces thyroxine, a hormone needed to maintain the high level of metabolism and heat production that is characteristic of mammals. Microscopic **parathyroid glands** are embedded within the tissue of the thyroid gland and secrete a hormone that regulates the calcium metabolism of the body. The human thyroid gland is located farther cranially on the ventral surface of the larynx.

B. BODY CAVITY

B.1. Opening the Body Cavity
To expose the organs in the thoracic and abdominal cavities, you have to make several incisions through the body wall (see Fig. 3-4). First make a shallow cut with a sharp scalpel through the skin just to the side of one of the longitudinal rows of nipples on the abdominal region. If you have already removed the skin, proceed with the next step. With a pair of forceps, lift the abdominal wall away from the underlying inner organs and make a small cut with a pair of scissors through the abdominal muscle layer and

the underlying whitish membrane (the peritoneum, see later). While being careful not to injure any organs or blood vessels leading to the body wall, extend this cut first caudally to the caudal end of the abdominal cavity (incision 2) and then cranially through the ribs of the thoracic cavity (incision 3). Next make corresponding parallel incisions through the body wall on the opposite side of the specimen (incisions 4 and 5).

As you spread apart the edges of the incisions in the thoracic region, you will discover a longitudinal membrane (the mediastinal septum, see later), which is attached along the strip of thoracic wall left between the two parallel incisions 3 and 5; do not damage it. You will also notice a transverse muscle wall, the **diaphragm**, attaching to the body wall and separating the thoracic cavity from the abdominal cavity.

With a pair of scissors, make a cut just cranially to and along the attachment of the diaphragm to the body wall all the way to the muscles of the back (incision 6). While doing so, you will cut through the caudal ribs of the thorax. Next make a corresponding cut on the other side of the specimen (incision 7), meanwhile leaving intact the central strip of body wall between incisions 3 and 5. Now, cut through one side of the body wall just caudally to and along the attachment of the diaphragm (incision 8). Finally, make a corresponding cut on the other side of the specimen (incision 9), but start this incision at the level of

incision 3 and cut across the midline. Be careful not to damage the umbilical blood vessel (the umbilical vein; see Exercise 4) that extends cranially from the umbilical cord to the liver.

If you are studying the muscular system, the third layer of the thoracic musculature (the transversus thoracis, Exercise 2) can now be seen on the inside of the thoracic wall near the sternum (see also Fig. 3-7).

B.2. Coelom

The thoracic and abdominal inner organs (or **viscera**) are situated within the body cavities, or **coelomic cavities**, which collectively are also called the **coelom**. Notice that the body cavities are completely lined with a shiny coelomic epithelium, the **serosa**, which also envelops all the viscera. In life, the serosa secretes a lubricating fluid, which facilitates the smooth gliding of the inner organs relative to one another and to the body wall. This possibility for relative motion is necessary because most inner organs change their shape and volume rhythmically (e.g., heart, lungs) or cyclically (e.g., stomach, intestine, uterus).

Cranial to the diaphragm, the coelom is divided into two lateral **pleural cavities** containing the lungs (see later) and, between the pleural cavities, a **pericardial cavity** containing the heart (Fig. 3-5). The serosa of the pleural cavities is called **pleura**. The **parietal pleura** lines the tho-

Body Cavity

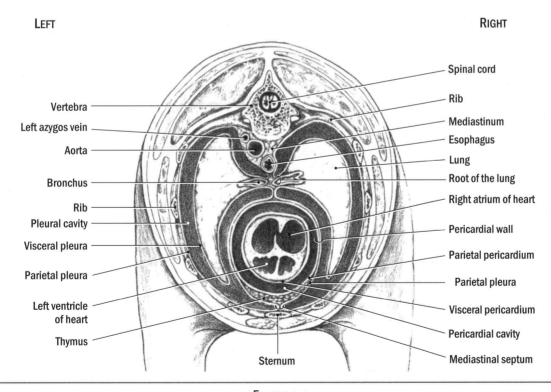

LEFT   RIGHT

Vertebra
Left azygos vein
Aorta
Bronchus
Rib
Pleural cavity
Visceral pleura
Parietal pleura
Left ventricle of heart
Thymus
Sternum

Spinal cord
Rib
Mediastinum
Esophagus
Lung
Root of the lung
Right atrium of heart
Pericardial wall
Parietal pericardium
Parietal pleura
Visceral pericardium
Pericardial cavity
Mediastinal septum

FIGURE 3-5
Diagrammatic cross section through the chest of a fetal pig at the level of the heart and viewed from the tail end.
The serosa is depicted as thin, white lines.

LEFT

RIGHT

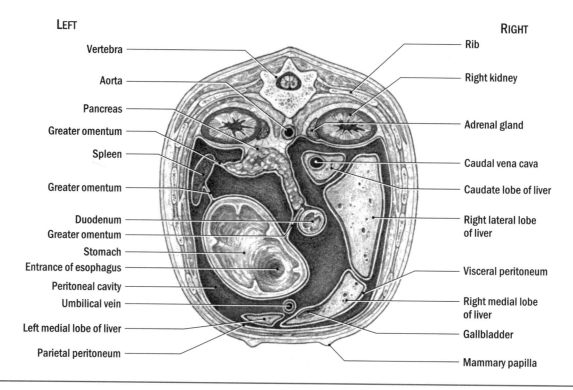

Vertebra — Rib

Aorta — Right kidney

Pancreas — Adrenal gland

Greater omentum

Spleen — Caudal vena cava

Greater omentum — Caudate lobe of liver

Duodenum — Right lateral lobe of liver
Greater omentum

Stomach

Entrance of esophagus — Visceral peritoneum

Peritoneal cavity — Right medial lobe of liver

Umbilical vein

Left medial lobe of liver — Gallbladder

Parietal peritoneum — Mammary papilla

FIGURE 3-6

Diagrammatic cross section through the abdomen of a fetal pig at the level of the stomach and viewed from the tail end. The serosa is depicted as thin, white lines.

racic wall and the membranous median wall between the paired pleural cavities. The **visceral pleura**, which is continuous with the parietal pleura, envelops the lungs. The space between the pleural cavities and the parietal pleurae forming their median walls, is called the **mediastinum**. This space is nearly filled with connective tissue, large blood vessels, the esophagus, the thymus, and the pericardial cavity (see Section C). Where the medial parietal pleurae of the paired pleural cavities touch each other directly, they form a **mediastinal septum**.

The pericardial cavity is lined by a serosa called **parietal pericardium**. The heart itself is enveloped by the **visceral pericardium**. The parietal pericardium is fused to the medial parietal pleura of the pleural cavities. The fused parietal pericardium and pleura form the **pericardial wall** of the pericardial cavity.

The part of the coelom caudal to the diaphragm is the **peritoneal cavity**, and the serosa in this region is called the **peritoneum**. **Parietal peritoneum** lines the body wall, and **visceral peritoneum** covers the abdominal organs (Fig. 3-6). The parietal peritoneum and the visceral peritoneum are continuous with each other and also form the **mesenteries**. Mesenteries are the thin membranes connecting the viscera to the body wall and to one another. They consist of two layers of serosa and enclose blood vessels and nerves that supply the viscera.

44

C. RESPIRATORY AND DIGESTIVE ORGANS OF THE THORAX

Look into the left pleural cavity of your specimen. On its medial wall, notice the heart lying in the pericardial cavity behind the transparent pericardial wall. Cranial to the heart, you can see the caudal end of the lobulated thymus, which you have already observed while dissecting the neck (Section A.2). The thymus lies within the mediastinum. Caudal to the heart, the medial wall is formed by the mediastinal septum, which is anchored to the sternum and the diaphragm. A whitish thread runs from the cranial end of the left pleural cavity across the thymus and the heart toward the diaphragm; this is the **left phrenic nerve**, a spinal nerve from the cervical region (see Exercise 7).

The lung is collapsed and does not fill the pleural cavity as it would in a live pig after birth. The lungs of a fetus, of course, have not yet been filled with air, nor have they begun to function, so they are much denser than lungs that have been inflated with air at birth. The left lung is divided into two lobes, an elongated **cranial lobe** and a larger **caudal lobe** (see Figs. 3-7 and 3-8). Notice that the two lobes merge medially at the level of the heart; this is the **root of the lung**. You may be able to see the large blood vessels radiating from the root into the lobes of the lung; these are the lung arteries. The root of the lung is also the part of the lung through which the **bronchi**, the terminal branches of the trachea, enter the lung, but you will not be able to see these until after the heart and its vessels have been dissected (see Exercise 4). The bronchi divide many times into smaller and smaller branches, the **bronchioles**.

With a pair of fine forceps, break through the visceral pleura covering the surface of the lung and dissect between the lobules of the lung tissue. You will see many blood vessels and bronchioles penetrating the lung tissue. The bronchioles end in microscopic **respiratory bronchioles**, which lead to thin-walled **alveolar sacs** and **alveoli** (Fig. 3-9A). These terminal lung structures are surrounded by capillaries and form a large, moist surface across which

gas exchange between the air and blood occurs. You can see many of these features in a histological slide preparation of an adult lung (Fig. 3-9B). The entire system of branching tubes from the trachea to the alveoli is known as the **respiratory tree**.

Now look into the right pleural cavity of your specimen. Find again the heart and the thymus behind the transparent medial wall. Caudal to the heart, you will find a very large, longitudinally oriented blood vessel, the **caudal vena cava** (see Exercise 4). A whitish thread runs from the cranial end of the right pleural cavity across the heart and along the caudal vena cava to the diaphragm. This is the **right phrenic nerve**. You can slide a blunt probe between the caudal vena cava and the lung lying dorsal to

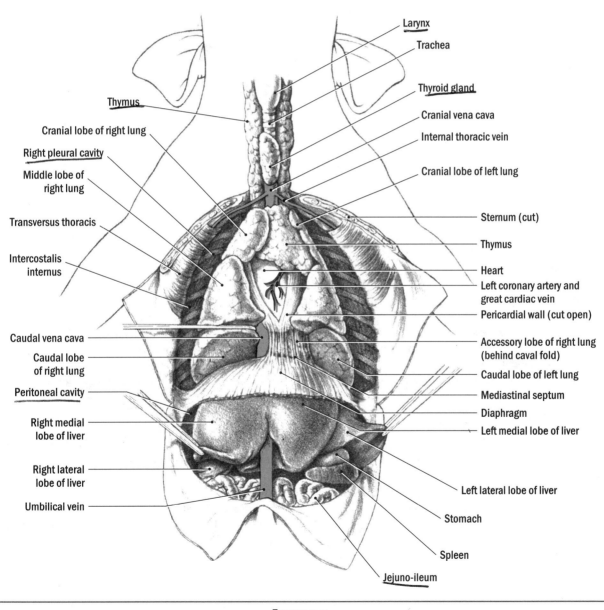

Respiratory and Digestive Organs of the Thorax

FIGURE 3-7
Ventral view of the organs in the thorax and cranial part of the peritoneal cavity of a fetal pig.

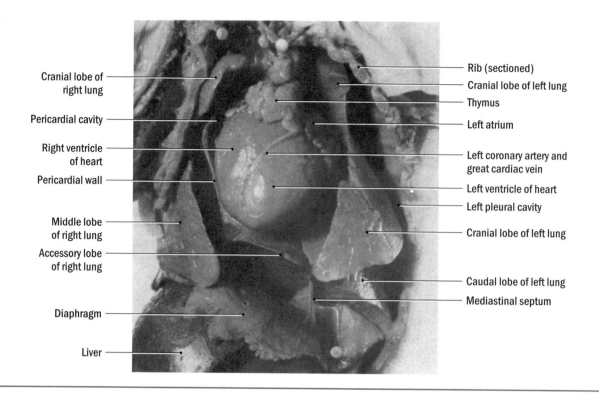

Cranial lobe of right lung

Pericardial cavity

Right ventricle of heart

Pericardial wall

Middle lobe of right lung

Accessory lobe of right lung

Diaphragm

Liver

Rib (sectioned)

Cranial lobe of left lung

Thymus

Left atrium

Left coronary artery and great cardiac vein

Left ventricle of heart

Left pleural cavity

Cranial lobe of left lung

Caudal lobe of left lung

Mediastinal septum

FIGURE 3-8
Photograph of a ventral view of the thoracic viscera of a fetal pig.
The caval fold is removed, and the caudal vena cava and the caudal lobe of the right lung do not show well in this view.

*Digestive
Organs of
the Abdomen*

it, and you will notice that the membrane anchoring the caudal vena cava to the sternum is a distinct fold of parietal pleura called the **caval fold**. Verify with your probe that the caval fold is separate from the mediastinal septum you saw while looking into the left pleural cavity.

The right lung consists of four lobes: a **cranial lobe**, a **medial lobe**, a large **caudal lobe**, and a small **accessory lobe** (Figs. 3-7 and 3-8). The accessory lobe crosses dorsally over the caudal vena cava and lies directly caudal to the heart within the recess formed between the caval fold and the mediastinal septum.

Return to the left pleural cavity. Pull the left lung ventrally and toward the middle of the body so that you can see the thoracic wall. Notice the straight, very large, round and whitish tube running along the midline of the thoracic wall dorsal to the root of the lung and heart. This is the **aorta**. At the level of the heart, it is crossed by a more flabby blood vessel, the **left azygos vein**, which receives branches from the intercostal muscles (see Exercise 4).

Carefully pick away the serosa cranial and dorsal to the crossing of the two blood vessels and you will discover a reddish, muscular but flabby tube, the **esophagus**. You may be able to introduce a blunt probe through the entrance of the esophagus in the pharynx (see Section A.2) by gently pushing it further and further without puncturing the esophageal wall. Caudal to the crossing of the two large blood vessels, the esophagus crosses to the right side

of the aorta and runs toward the diaphragm. It is accompanied by branches of the **vagus**, a cranial nerve that supplies the viscera (see Exercise 7).

Follow the esophagus to the diaphragm. Note that the periphery of the diaphragm is muscular, whereas its center is translucent and tendinous. Make a radial cut through the left side of the diaphragm toward the esophagus. The diaphragm surrounding the esophagus consists of muscle tissue that forms a sling attaching the esophagus to the dorsal wall of the thoracic and abdominal cavities.

D. DIGESTIVE ORGANS OF THE ABDOMEN

D.1. Stomach
The most conspicuous organ in the abdominal cavity is the **liver** (*hepar*), which fits under the dome of the diaphragm. The umbilical vein, which runs from the umbilical cord toward the liver and which you saw while opening the abdominal cavity (see Section B.1), divides the liver into a right and a left half. As you lift the **left** and **right medial lobes of the liver**, you will discover the **left** and **right lateral lobes of the liver** (see Figs. 3-6, 3-7 and 3-10). If you now lift the right lateral lobe of the liver, you will also discover the **caudate lobe of the liver**, which is partly

hidden under coils of intestine. The large caudal vena cava (see Exercise 4) passes through the caudate lobe of the liver.

The liver is anchored to the ventral abdominal wall by the umbilical vein and its associated peritoneum called the **falciform ligament**. The liver is also attached to the diaphragm by the **coronary ligament**, which is essentially peritoneum bridging the gap between the liver and the diaphragm. It is seen best on the left lateral lobe of the liver. To view the coronary ligament, you may have to bisect the umbilical vein and the falciform ligament.

Lift the left half of the liver to reveal the **stomach** (*ventriculus*) (see Fig. 3-10). Find the esophagus again and

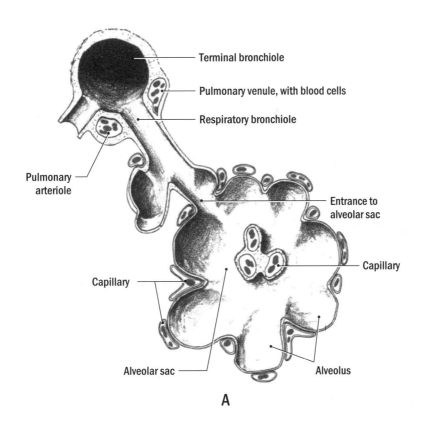

Terminal bronchiole

Pulmonary venule, with blood cells

Respiratory bronchiole

Pulmonary arteriole

Entrance to alveolar sac

Capillary

Capillary

Alveolar sac

Alveolus

Digestive Organs of the Abdomen

A

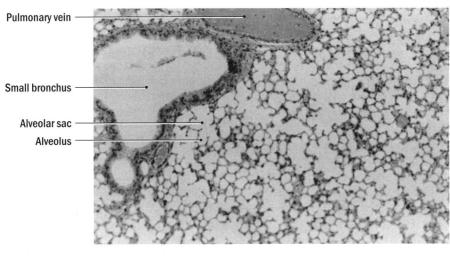

Pulmonary vein

Small bronchus

Alveolar sac

Alveolus

B

FIGURE 3-9
The mammalian lung.
(A) Diagram of a terminal portion of the respiratory tree of a mammal.
(B) Photomicrograph through a portion of the lung of a mouse.

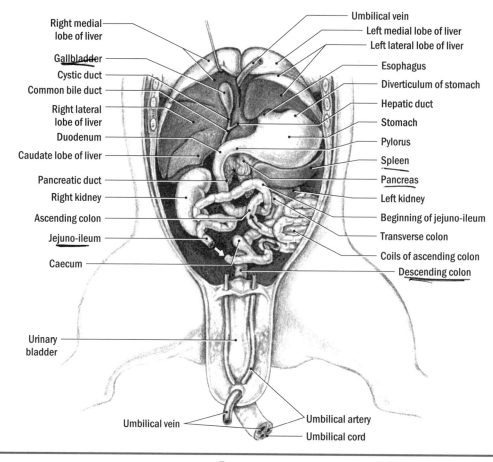

Right medial lobe of liver
Gallbladder
Cystic duct
Common bile duct
Right lateral lobe of liver
Duodenum
Caudate lobe of liver
Pancreatic duct
Right kidney
Ascending colon
Jejuno-ileum
Caecum

Urinary bladder

Umbilical vein

Umbilical vein
Left medial lobe of liver
Left lateral lobe of liver
Esophagus
Diverticulum of stomach
Hepatic duct
Stomach
Pylorus
Spleen
Pancreas
Left kidney
Beginning of jejuno-ileum
Transverse colon
Coils of ascending colon
Descending colon

Umbilical artery
Umbilical cord

*Digestive
Organs of
the Abdomen*

FIGURE 3-10
Ventral view of the abdominal viscera of a fetal pig. The mesenteries have been removed.
The liver has been lifted and pulled cranially, and most of the jejuno-ileum has been removed (see white arrow).

notice that it opens into the stomach immediately after it has passed through the diaphragm. Pull it cranially to get a better view of the stomach, which lies caudal to the liver on the left-hand side of the body (Fig. 3-10). The stomach is a large, saccular, somewhat J-shaped organ in which food accumulates before continuing in small portions on its way through the intestine. In the stomach, salivary enzymes continue to act on the food for a while, but the gastric secretion containing hydrochloric acid and the proteolytic enzyme pepsin soon stop this activity and begin to break down proteins and, in particular, the collagen of the connective tissue in meat. (Remember that the pig is omnivorous.)

The stomach of the pig can be subdivided into four regions. Adjacent to the esophageal entrance is its **cardiac region**. The blind sac to the left of the esophagus is the **diverticulum**, which corresponds to the fundus in other mammals with a simple stomach, such as human beings. The main saccular part of the stomach is its **body**, and the thick-walled portion near the entrance to the intestine on the right side of the body is its **pylorus**. The short cranial edge of the stomach between the entrance of the esophagus and the pylorus is its **lesser curvature**; the longer caudal edge is its **greater curvature**.

A mesentery known as the **lesser omentum** extends from the lesser curvature of the stomach and small intestine to the liver. Another mesentery, the **greater omentum**, forms an empty sac that extends caudally from the greater curvature of the stomach and then curves dorsally and folds back cranially to attach to the duodenum (see later) and the dorsal body wall. Between the two layers of serosa forming this mesentery, fat is deposited along the blood and lymphatic vessels that form a network within the greater omentum.

The dark elongated **spleen** is enclosed between the two layers of serosa of the portion of the greater omentum near the left side of the stomach. The spleen in a fetus is part of the blood-forming (hematopoietic) tissues of the body and produces red blood cells. In an adult pig, the spleen produces lymphocytes and is part of the immune system. It also captures and breaks down senescent red and white blood cells. The end products of this breakdown reach the liver, where they are eliminated with the bile (e.g., bile pigments from the breakdown of hemoglobin).

Cut open the stomach and wash it out. The greenish slimy substance found here and elsewhere in the digestive tract is **meconium**. It consists of bile-stained mucus and sloughed-off epithelial cells of the skin and lining of the digestive tract. During the fetal period, some of this material is discharged into the amniotic fluid surrounding the embryo, from which it reenters the digestive tract when the fetus swallows amniotic fluid. It is all discharged in the first, greenish bowel movements of the newborn.

Notice that the stomach wall is particularly thick at the pylorus and that the lumen of the pylorus is partly occluded by a fold. This region is called the **pyloric canal**, and its thick muscular wall forms the **pyloric sphincter**, which keeps food in the stomach until it is broken down sufficiently to be handled by the intestine and then releases the food in small portions into the intestine. A less conspicuous **cardiac valve** between the stomach and esophagus normally prevents the highly acidic stomach contents from backing up into the esophagus. Near the entrance of the esophagus on the left side of the body, you will be able to verify that the diverticulum is a side chamber of the stomach, from which it is demarcated by a ring-like fold.

D.2. Small Intestine and Digestive Glands

The first part of the small intestine is the duodenum. Trace it caudally from the pylorus. To do so, you will have to lift the coils of the rest of the small intestine, the **jejuno-ileum**. Notice that the duodenum is accompanied by some lobulated tissue (the pancreas, see later) as it passes over the caudate lobe of the liver. You can introduce a blunt probe between the duodenum and the caudate lobe of the liver with the caudal vena cava. This opening is the **epiploic foramen**, which is relatively larger in the fetus than in the adult. If you direct your probe cranially and dorsally to the stomach, the probe will be seen emerging dorsal to the transparent lesser omentum (see Section D.1). If you direct the probe toward the left of your specimen, the probe will enter the mesenteric sac formed by the greater omentum.

Beyond the epiploic foramen, the duodenum is anchored to the body wall and connected to the mass of intestinal coils by a mesentery. With a pair of fine forceps, carefully tear this mesentery as you continue to follow the duodenum, but do not break any blood vessels or separate the pancreatic tissue from the duodenum. The duodenum crosses to the left side of the body by passing dorsal to the coils of the jejuno-ileum. It finally turns cranially until it reaches a larger, darker, transversely oriented segment of the intestine that is attached to the dorsal body wall (the transverse colon, see later and Fig. 3-10).

The jejuno-ileum starts where the duodenum turns to the right of the specimen along the transverse colon. Follow the convoluted loops of the jejuno-ileum and notice that they are gathered by a central mesentery, which is called simply **the mesentery**. This arrangement prevents the very long small intestine from getting entangled. The mesentery contains many blood and lymphatic vessels.

To see the extensive **pancreas**, lift and turn the stomach in such a manner that its greater curvature points cranially. Find again the connection between the pancreas and the duodenum near the pylorus. Carefully separate the transverse colon (see earlier) and push the transverse colon caudally. Notice that the pancreas is connected to the greater omentum and that the greater omentum is a continuous sheet of mesentery between the greater curvature of the stomach and the dorsal body wall. The duodenum, pancreas, and spleen (see earlier) are built into this greater omentum (see Fig. 3-6).

The pancreas secretes enzymes that act on all major categories of food (carbohydrates, proteins, lipids, and nucleic acids). It also secretes bicarbonate ions, which neutralize the highly acidic food coming from the stomach, because the pancreatic enzymes can act only in a neutral milieu. The pancreatic enzymes and the bicarbonate ions enter the duodenum through a minute **pancreatic duct**, which can be found by carefully removing the peritoneum from the caudal end of a little tongue of pancreatic tissue that follows along the descending part of the duodenum.

The pancreas also contains microscopic patches of endocrine tissue, the **islets of Langerhans**, which produce insulin and glucagon. These two hormones are released directly into the circulatory system and regulate the blood glucose levels and the cellular metabolism of glucose.

Find the liver again and lift it to see its dorsal surface. Notice the **gallbladder** (*vesica fellea*) embedded in the caudal face of the right medial lobe of the liver just to the right of the point at which the umbilical vein enters the liver (see Figs. 3-6 and 3-10). Carefully remove the peritoneum from the base of the gallbladder and from the lesser omentum. A **cystic duct**, accompanied by a small artery, leads out of the gallbladder and joins several **hepatic ducts** (one of which is rather prominent) from the liver to form a **common bile duct** (*choledochal duct*), which enters the duodenum just beyond the pylorus. As you do this dissection, you may notice the **hepatic artery** and the **hepatic portal vein** in the lesser omentum deep to the common bile duct (see Exercise 4).

At the end of the bile duct, there is a sphincter that is usually closed. Therefore, the bile, which is secreted by the liver, accumulates in the gallbladder. When food enters the duodenum, the sphincter at the end of the common bile duct relaxes and bile is discharged into the intestine. There are no digestive enzymes in bile, but it is very alkaline and helps to neutralize the acidic food coming from the stomach. Bile contains bile salts that emulsify fat within the generally watery mixture of food and secretions from the digestive tract. Emulsification of fats facilitates the enzymatic breakdown of fat by the water-soluble lipase from the pancreas. In this manner, bile salts also promote the absorption of fats and fat-soluble vitamins by the small intestine. Bile also contains bile pigments as a waste product from the breakdown of hemoglobin in the liver; they give the bile and feces their characteristic color. The liver not only produces bile but also plays a crucial role in the storage and metabolism of nutrients, as well as in the conversion of toxic by-products from the metabolism of

Digestive Organs of the Abdomen

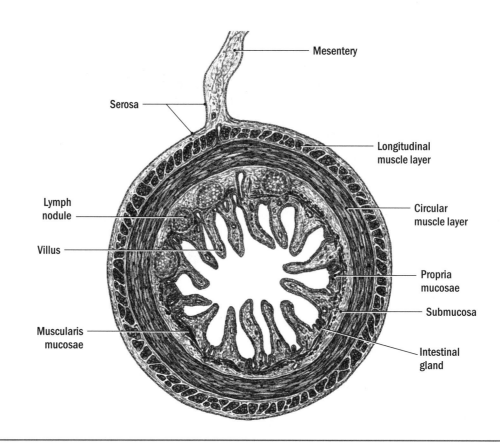

Serosa

Mesentery

Longitudinal
muscle layer

Circular
muscle layer

Lymph
nodule

Villus

Propria
mucosae

Submucosa

Muscularis
mucosae

Intestinal
gland

FIGURE 3-11
Diagram of a microscopic transverse section through the ileum of a cat.

proteins and other nitrogenous compounds (e.g., ammonia) into less toxic substances (e.g., urea).

Cut open a part of the small intestine, wash it out, and notice the velvety texture of its lining, which is generated by minute projections, the **villi**. The openings of microscopic intestinal glands lie between the bases of the villi, and the enzymes secreted by these glands, together with those of the pancreas, complete the breakdown of food, except for the fiber called cellulose. The absorptive surface area of the small intestine is greatly increased by the villi.

D.3. Large Intestine
Follow again the coils of the small intestine until it eventually enters the large intestine, or **colon**, at the **caeco-colic junction**. The **caecum** of the pig is a short, blind sac that is connected to the jejuno-ileum by a mesentery. The caecum continues into the colon, which has a much larger diameter than the small intestine. Cut open the colon opposite the entrance of the small intestine and notice the **ileal ostium**, the opening of the small intestine into the large intestine. The ileal ostium is surrounded by the **ileal sphincter**, which prevents a reflux of intestinal contents from the large intestine into the small intestine. The ileal

sphincter forms an **ileal papilla**, which projects into the large intestine. Notice also that the internal surfaces of the caecum and the colon are smooth and devoid of villi. The intestinal epithelial lining contains numerous goblet cells, which produce mucus to lubricate the passage of the food through the colon.

Most of the colon of the pig forms a tightly coiled mass with a unique pattern. It spirals in centrifugal loops on the outside of the coil, reverses at the apex of the coil, and spirals back in centripetal turns inside the coil. The centripetal loops can be seen only after the centrifugal loops have been separated from one another by breaking the peritoneum. This coiled part of the colon in the pig corresponds to the **ascending colon** of human beings and other mammals. As the colon emerges from the coiled mass, it continues as the **transverse colon**, which was seen earlier, and finally turns caudally as the **descending colon** (see Fig. 3-10).

The descending colon is anchored to the dorsal body wall by a very short mesentery, the **mesocolon**. The colon continues as the **rectum** as it enters the **pelvic cavity**, which is encased by the pelvic girdle. The rectum, or terminal part of the large intestine, will be seen in a later dissection (see Exercise 5). It opens on the body surface through the **anus**.

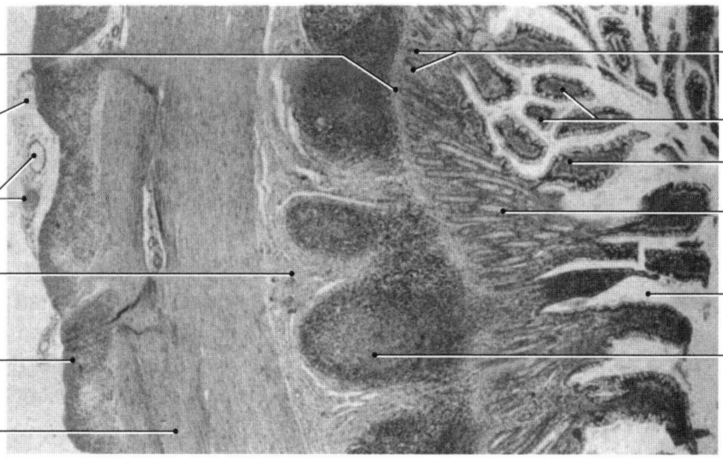

Muscularis mucosae
Serosa
Small artery and vein
Submucosa
Longitudinal muscle layer
Circular muscle layer

Intestinal gland
Villi
Villus
Intestinal gland
Intestinal lumen
Lymph nodule

Propria mucosae

FIGURE 3-12
Photomicrograph of a transverse section through the wall of the ileum of a cat.

Certain salts, vitamins, and a considerable amount of water are absorbed from the digestive residues in the colon. Much of the water comes from the digestive secretions released into the mouth, stomach, and small intestine. In addition, symbiotic bacteria, which reproduce in abundance in the colon, synthesize most of the vitamin K needed by the liver of mammals for the production of certain blood-clotting factors. The symbiotic bacteria also secrete cellulases, which are enzymes that cannot be produced by any multicellular organisms, including mammals. Cellulases break down cellulose, or fiber, into monosaccharides, such as glucose, which, in turn are fermented into acetic acid and other volatile fatty acids within the colon. These volatile fatty acids are absorbed by the colon. Plant material, which contains a great deal of cellulose, is an important component of a pig's diet and is digested mostly in the colon. In rodents and horses, whose diets consist completely of plant material, the caecum is very long and hosts the microorganisms responsible for the digestion of cellulose. Human beings possess only a very short caecum with a vermiform appendix.

Undigested residues, disintegrated bile pigments, and many bacteria are finally discharged as the feces.

E. MICROSCOPIC STRUCTURE OF THE SMALL INTESTINE

Study a histological slide preparation of a cross section through the small intestine of a mammal, and compare it with Figs. 3-11, 3-12, and 3-13. Beginning at the outside and progressing toward the lumen, note the following layers: (1) serosa, (2) longitudinal muscle layer, (3) circular muscle layer, (4) submucosa, (5) mucosa.

The **serosa** is a thin layer of simple squamous epithelium (the visceral peritoneum) supported by connective tissue, which contains a few blood vessels. About all that can be seen of the epithelial cells are their flattened nuclei. The attachment of the mesentery to the intestine will be seen in some slides.

The **longitudinal muscle layer** is composed of **smooth muscle** fibers that spiral around the intestine in a plane close to its longitudinal axis. Because these are overlapping, spindle-shaped cells being viewed primarily in cross section, their diameters vary considerably, as can be seen under high magnification. Smooth muscle fibers in the **circular muscle layer** form a tight spiral whose coils lie close to the transverse plane of the intestine and, therefore, are seen in longitudinal sections. Notice that, under high magnification, each of these muscle fibers can be seen to contain a single elongate, centrally located nucleus and many longitudinal, nonstriated myofibrils.

These two muscle layers, longitudinal and circular, function as antagonists to each other; they produce the churning movements that mix food and enzymes and the peristaltic movements that move the food through the intestine. They are innervated by the autonomic nervous system, of which the parasympathetic provides activating stimuli, while the sympathetic provides inhibiting stimuli. The cell bodies of the postganglionic parasympathetic neurons (see Exercise 7) are concentrated in lighter stained **intramural ganglia** within the wall of the digestive tract, either between the longitudinal and cir-

Microscopic Structure of the Small Intestine

51

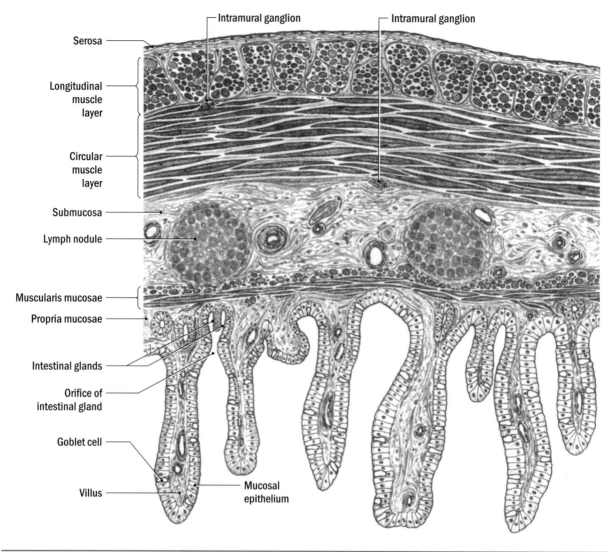

- Intramural ganglion

- Intramural ganglion

Serosa

Longitudinal
muscle
layer

Circular
muscle
layer

Submucosa

Lymph nodule

Muscularis mucosae

Propria mucosae

Intestinal glands

Orifice of
intestinal gland

*Microscopic
Structure of the
Small Intestine*

Goblet cell

Villus

Mucosal
epithelium

FIGURE 3-13
Diagram of an enlarged microscopic transverse section through the wall of the ileum of a mammal.

cular muscle layers or between the circular muscle layer and the submucosa.

The submucosa consists of loose fibrous connective tissue containing many blood vessels. If your histological slide was taken from near the colic end of the small intestine, large, deep-staining masses of lymphocytes called **lymph nodules** will be seen along one side of the intestine, lying in the submucosa and extending into the mucosa. As already mentioned, T-lymphocytes mature in the thymus. It is probable that a second type, B-lymphocytes, mature in lymphoid tissue in the wall of the digestive tract before they are distributed to other lymphoid tissues. On exposure to foreign antigens, primarily of bacterial origin, certain B-lymphocytes transform themselves into antibody-producing cells.

The **mucosa** is the most complex layer of the intestine and can be subdivided into three layers. The **muscularis mucosae** is a thin layer of longitudinal and circular smooth muscle fibers and lies internally next to the submucosa. The **propria mucosae**, the layer of connective tissue next to the muscularis mucosae, contains blood and lymphatic vessels. The **mucosal epithelium** lines the intestinal lumen; it is a simple columnar epithelium and contains many lightly stained **goblet cells**. These cells secrete mucus, which lubricates the food within the intestinal lumen and helps protect the delicate intestinal lining, especially from the effects of proteolytic enzymes.

The numerous fingerlike projections that extend into the lumen of the intestine are called **villi**. The core of each villus is formed by the propria mucosae and some muscle fibers that extend from the muscularis mucosae. The latter convey motility to the villi. The villi and the ultramicroscopic **microvilli** on the surface of the epithelial cells greatly increase the absorptive surface area of the small intestine.

Between the villi, the mucosal epithelium dips deeply into the propria mucosae and forms the tubular **intestinal glands**. As a result of the three-dimensional arrangement of these glands, your histological slide preparation may show the intestinal glands and villi only in cross or oblique sections. The intestinal glands and the columnar cells of the mucosal epithelium secrete intestinal enzymes, some of which are released into and act within the lumen of the intestine, whereas others remain associated with the microvilli at the surface of the epithelial cells.

Microscopic Structure of the Small Intestine

Circulatory System

THE CIRCULATORY SYSTEM is the transport system of the body. In an adult animal, it carries substances absorbed by the digestive tract and oxygen absorbed from the lungs to all body cells, and it carries waste products of cellular metabolism to organs of waste removal. Carbon dioxide is eliminated by the lungs and nitrogenous wastes by the kidneys. In a fetus, all nutritional, respiratory, and excretory exchanges take place through the placenta. In both adult and fetal animals, substances enter and leave the body and its cells by diffusion, supplemented in some cases by active transport. But diffusion is effective only over short distances; thus, in all animals that are larger than a few millimeters, a circulatory system is necessary to provide a mechanism for the bulk flow of materials between sites of entry into the body, sites of consumption (the cells), and sites of waste removal. In essence, a circulatory system allows animals to grow larger than a few millimeters and to lead active lives that require a high level of metabolism. The circulatory system also helps to distribute heat throughout the body, transports hormones, helps to maintain a constant internal environment, and defends the body against disease organisms by transporting antibodies and immunoactive blood cells to infected body parts.

A. ADULT CIRCULATION

Although you will be studying a fetus, the blood vessels you will expose will mean more to you if, before dissecting them, you understand the basic pattern of circulation in an adult mammal. Blood is transported in **arteries** away from the heart to minute, thin-walled **capillaries** in the tissues, where exchanges of water, nutrients, gases, and waste products between the blood and interstitial (extracellular) fluid, which bathes the cells, take place. The capillary beds are drained primarily by **veins**, which return blood to the heart, but some surplus interstitial fluid, and a few smaller protein molecules that seep out of the capillaries, return to the larger veins by first entering **lymphatic vessels**. Lymphatic vessels are difficult to see in gross dissections, although some of the lymph nodes that lie along their course will be noticed. **Lymph nodes** are sites for the production and storage of certain white blood cells (lymphocytes), which play a critical role in the body's immune defense.

Unlike the open circulation of many invertebrates, in which all components of the blood pass freely among the cells of the body, the circulatory system of vertebrates is a closed system. Most components of the blood remain in the blood vessels and, for the most part, only small molecules are exchanged between the capillaries and the interstitial fluid. Although an open circulatory system is well suited to the mode of life of many invertebrates, a closed circulatory system allows for higher blood pressures and a more rapid distribution of materials; the closed circulatory system also allows, by dilation and constriction of selective blood vessels, for control over the amount of blood going to different regions of the body. Active organs receive more blood than less active ones.

In an adult mammal (Fig. 4-1), blood that is oxygen-depleted and carbon dioxide-enriched is returned from most of the body to the **right atrium** of the heart by two major veins and their tributaries. The **cranial vena cava** drains the head, neck, and arms; the **caudal vena cava** drains the caudal parts and inner organs of the body. It is important to recognize that the blood that is drained from the stomach and intestine is led first by the **hepatic portal vein** to capillary-like spaces within the liver, the **hepatic sinusoids**. Many metabolic conversions of absorbed materials occur in the liver, from which the blood drains into the caudal vena cava through several **hepatic veins**.

As oxygen-depleted and carbon dioxide-enriched blood enters the right atrium, blood that has acquired oxygen and released carbon dioxide in the capillaries of the lungs returns to the **left atrium** of the heart in the **pulmonary veins**. (Notice that veins are blood vessels that lead to the heart, regardless of the oxygen content of their blood. Similarly, arteries always lead away from the heart.) The right and left atria contract simultaneously and discharge their blood into the right and left ventricles, which

are expanding by elastic recoil after their previous contractions. When the thick, muscular walls of the ventricles contract, the increasing pressure closes the **atrioventricular valves**, and blood is driven with considerable force into the two large arteries leaving the heart. The closure of valves at the base of these arteries prevents a backflow of blood into the ventricles as they refill from the atria. The **pulmonary trunk** leaves the **right ventricle** and soon branches into a pair of **pulmonary arteries**, which carry the oxygen-depleted blood to the lungs, where gas exchange takes place in an adult mammal. The **aorta**, which carries oxygen-rich blood in an adult mammal, leaves the **left ventricle**, curves to the left side of the body, and runs toward the pelvis. On its way, it gives off numerous branches that deliver oxygen-rich blood to all parts of the body.

Blood flowing through the heart does not supply the heart musculature itself with needed oxygen and nutrients. A separate system of **coronary arteries** leaves the base of the aorta and leads to capillary beds in the heart wall. **Cardiac veins** drain the heart wall and return oxygen-depleted blood via the coronary sinus to the right atrium.

Remember that you are dissecting a fetal specimen. Some aspects of the circulatory pattern will be different from that in an adult because the placenta is the site for the exchange of gases, nutrients, and waste products. The differences between fetal and adult circulatory patterns, and the changes that occur at birth, will be summarized after you have seen the major blood vessels (see Section F).

The arteries and major veins should have been injected in your specimen, but valves in the veins sometimes prevent the thorough penetration of the material with which they have been injected. If a vein is uninjected, it will appear as a thin-walled, translucent tube in which a bit of blood or injection fluid can be seen. Because most veins lie beside corresponding arteries, it is convenient to study the arteries and veins together in many regions of the body. Do not be surprised to find variation among individual specimens in the locations at which smaller blood vessels join the main ones. Often, the only way to be certain of the identity of a blood vessel is to trace it to the organ it supplies or drains.

B. THE HEART AND ITS GREAT BLOOD VESSELS

Spread apart the rib cage after you have opened it (see Exercise 3, Fig. 3-4). Carefully cut the mediastinal septum and pericardial wall from their attachment on the inner surface of the sternum. Do not injure the **phrenic nerves**, which supply the diaphragm, or any blood vessels as you do so. You may need to detach the thymus from the heart and sternum and pick it carefully away from the base of

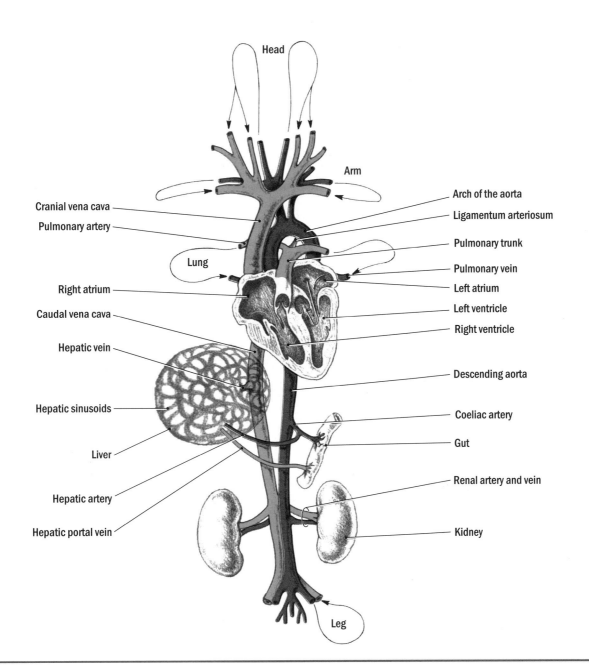

Head

Arm

Cranial vena cava

Pulmonary artery

Lung

Right atrium

Caudal vena cava

Hepatic vein

Hepatic sinusoids

Liver

Hepatic artery

Hepatic portal vein

Arch of the aorta

Ligamentum arteriosum

Pulmonary trunk

Pulmonary vein

Left atrium

Left ventricle

Right ventricle

Descending aorta

Coeliac artery

Gut

Renal artery and vein

Kidney

Leg

*The Heart
and Its Great
Blood Vessels*

FIGURE 4-1
Diagrammatic ventral view of the course of blood flow in an adult mammal.
Red indicates oxygen-rich blood, and blue indicates oxygen-depleted blood.

the neck to have an unobstructed view of the blood vessels in this region. Notice a pair of veins accompanied by small arteries extending from vessels at the base of the neck onto the ventral side of the thorax, passing on each side of the sternum. These are the **internal thoracic vessels**. Try not to break them.

The heart (Fig. 4-2) consists of the **right** and the **left ventricles**, which can often be identified externally on the ventral surface of the heart by a groove that separates them and contains a prominent **left coronary artery** and **great cardiac vein.** The **right atrium** and the **left atrium**, each with an ear-shaped, dark-colored **auricle**, lie on each side of the cranial end of the heart and are conspicuous in a ventral view.

Several veins draining the head, neck, and arms converge at the base of the neck to form the **cranial**, or **57**

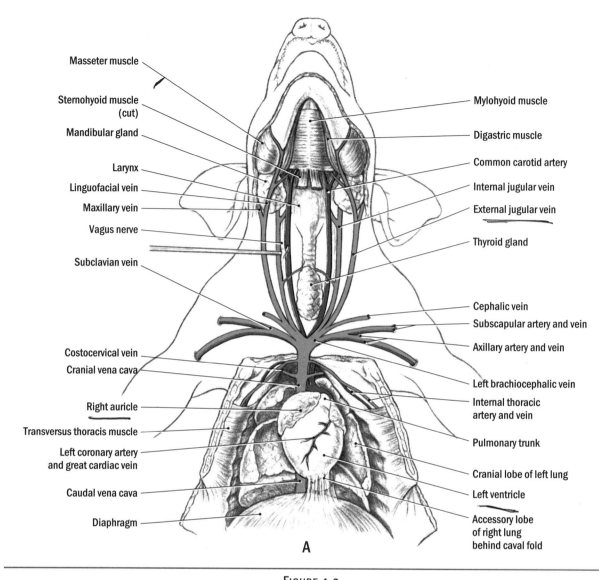

Masseter muscle

Sternohyoid muscle (cut)

Mandibular gland

Larynx

Linguofacial vein

Maxillary vein

Vagus nerve

Subclavian vein

Costocervical vein

Cranial vena cava

Right auricle

Transversus thoracis muscle

Left coronary artery and great cardiac vein

Caudal vena cava

Diaphragm

Mylohyoid muscle

Digastric muscle

Common carotid artery

Internal jugular vein

External jugular vein

Thyroid gland

Cephalic vein

Subscapular artery and vein

Axillary artery and vein

Left brachiocephalic vein

Internal thoracic artery and vein

Pulmonary trunk

Cranial lobe of left lung

Left ventricle

Accessory lobe of right lung behind caval fold

A

The Heart and Its Great Blood Vessels

FIGURE 4-2
Ventral views of the veins and arteries of the neck and head of a fetal pig.
(A) Overall view after removal of the thymus and superficial neck muscles. Veins are colored blue and arteries red to aid in distinguishing them, but in a fetus many of these blood vessels carry a mixture of oxygen-rich and oxygen-depleted blood. The specimen shown lacked the second subclavian vein.

superior, **vena cava**, which enters the right atrium. The **caudal**, or **inferior**, **vena cava** perforates the diaphragm and enters the right atrium caudally. Its tributaries drain the caudal parts of the body, and, in the fetus, the placenta as well. The **left** and **right phrenic nerves** lie on each side of the cranial vena cava. The right phrenic nerve extends along the caudal vena cava.

Blood from the right atrium flows into the right ventricle, which, in an adult, pumps it to the lungs through the **pulmonary trunk**. The pulmonary trunk, a very large blood vessel, leaves the cranial end of the right ventricle and curves dorsally between the auricles (Fig. 4-3). The **aorta**, another very large blood vessel, leaves the cranial end of the left ventricle and lies dorsal to the pulmonary trunk. With a pair of forceps, carefully pick away connective tissue between the

pulmonary trunk and the aorta. Then pick away connective tissue between the pulmonary trunk and the left atrium to find where the pulmonary trunk bifurcates into the two **pulmonary arteries**, which continue toward the lungs.

In the fetus, a large branch of the pulmonary trunk continues dorsally and laterally as a **ductus arteriosus**, which joins the aorta. Because the lungs are not functioning for respiration in a fetus, most of the fetal blood from the right ventricle bypasses the lungs by flowing through the ductus arteriosus into the aorta. Some fetal blood bypasses the lungs by flowing directly from the right atrium through an opening in the interatrial wall (the foramen ovale) to the left atrium (Section F).

Blood picks up oxygen and releases carbon dioxide in the lungs of an adult and returns to the left atrium in

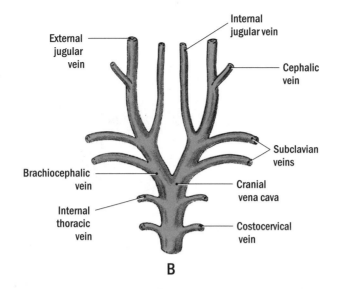

External
jugular
vein

Internal
jugular vein

Cephalic
vein

Subclavian
veins

Brachiocephalic
vein

Cranial
vena cava

Internal
thoracic
vein

Costocervical
vein

B

FIGURE 4-2 CONTINUED
Ventral views of the veins and arteries of the neck and head of a fetal pig.
(B) Diagram of the major veins of a specimen with two subclavian veins on each side. (B, modified after Nickel, R.,
Schummer, A., and Seiferle, E. *The Viscera of Domestic Animals*. New York: Springer-Verlag, 1979.)

*Veins and
Arteries of the
Neck, Head,
and Thorax*

the **pulmonary veins**. A **left azygos vein**, which receives much of the drainage of the thoracic wall, lies lateral to the pulmonary arteries. We will return to it later. One of the left pulmonary veins leaves the base of the cranial lobe of the left lung and can be seen passing between the left azygos vein and the left pulmonary artery. The remaining pulmonary veins and their entrance into the dorsal surface of the left atrium will be seen when the heart has been removed (Fig. 4-8, Section E).

The left atrium receives blood from the lungs in an adult, but most of the blood in the left atrium of a fetus has come from the right atrium through the interatrial foramen. In either case, blood from the left atrium enters the left ventricle, which pumps it into the aorta. As you have seen, the aorta leaves the heart dorsal to the pulmonary trunk and then emerges to the right of the pulmonary trunk before it arches to the left side of the body as the **arch of the aorta** and turns caudally as the **descending aorta** (Figs. 4-1 and 4-3).

C. VEINS AND ARTERIES OF THE NECK, HEAD, AND THORAX

Veins in the neck lie superficial to the arteries and should be studied first. Because there is some mixing of blood in the heart of a fetus, the color of the injection mass in a fetal specimen (usually blue for veins and red for arteries) is not a reliable indicator of the identity of a blood vessel. In general, veins have thinner walls and a flattened cross section, whereas arteries have thicker walls and a rounded cross section. Some of the veins may have been partially destroyed when you studied muscles and organs in the neck during Exercises 2 and 3, but examine their pattern in Figure 4-2A and find as many as you can. Surrounding connective tissue must be picked away carefully. Many veins lie beside accompanying arteries, which may be more easily seen.

A pair of relatively large **external jugular veins** lie superficially on the lateroventral surface of the neck. One of them was used to inject the blue mass into the venous system. A **cephalic vein** lies just beneath the skin on the side of the arm and typically enters the external jugular vein near its base. It often is not, or is only partially, injected. These veins drain most of the head and neck and part of the shoulder region.

A pair of deeper and more medial **internal jugular veins** lie close to the trachea just lateral to the common carotid arteries (see later). The internal jugular veins drain part of the inside of the skull. External and internal jugular veins on each side of the body usually join each other and very soon unite with *one* of the **subclavian veins** coming from the shoulder and arm. (Note that the pig is unusual among mammals in that it often has two subclavian veins on each side of the body, one lying dorsal and one ventral to the subclavian artery, see later.) The confluence of these vessels forms on each side of the body a short **brachiocephalic vein**, which receives the second subclavian vein (Fig. 4-2B). Sometimes the external jugular, internal jugular, and one subclavian vein enter the brachiocephalic vein together. The brachiocephalic veins turn deeply and unite with each other to form the cranial vena cava, which you have already seen entering the heart.

The two subclavian veins on each side, accompanied by the subclavian artery, extend laterally deep into the

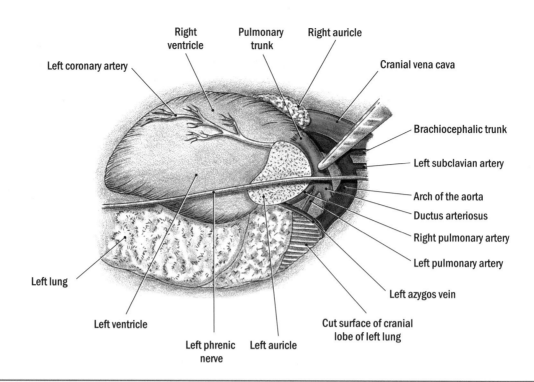

Right
ventricle

Pulmonary
trunk

Right auricle

Left coronary artery

Cranial vena cava

Brachiocephalic trunk

Left subclavian artery

Arch of the aorta

Ductus arteriosus

Right pulmonary artery

Left pulmonary artery

Left azygos vein

Left lung

Left ventricle

Left phrenic
nerve

Left auricle

Cut surface of cranial
lobe of left lung

FIGURE 4-3
Ventrolateral view of a dissection of several great vessels entering and leaving the heart in a fetal pig.

*Veins and
Arteries of the
Neck, Head,
and Thorax*

arm, but different portions of the veins are given different names, according to the regions through which they pass: **subclavian veins** (lying beneath the clavicle in a human being), **axillary veins** (passing through the axilla, or arm-pit), and **brachial vein** (entering the arm). Sometimes a subclavian vein can be identified most certainly by tracing it into the upper arm; but if you do this, you must separate veins, arteries, and nerves in the region very carefully. Near the shoulder joint, the two axillary veins are interconnected by anastamosing branches. At the shoulder joint, both axillary veins curve distally and continue as a single brachial vein. You may also notice a **subscapular vein** (Fig. 4-2A), which drains the medial side of the scapula and enters either the subclavian or axillary vein.

If your specimen is well injected, certain smaller tributaries of the cranial vena cava can also be found. You may have already noticed the pair of **internal thoracic veins** entering the ventral surface of the cranial vena cava shortly caudal to the point where the vena cava is formed by the confluence of the brachiocephalic veins. They extend along the lateral border of the internal surface of the sternum deep to a thin sheet of muscle (the transversus thoracis, Exercise 3), accompanied by internal thoracic arteries, and help to drain the chest wall. A pair of **costocervical veins** can be found closer to the heart, but more deeply, because they enter the dorsal surface of the cranial vena cava. They drain parts of the neck, back, and cranial intercostal spaces, and they are accompanied by costocervical arteries.

Turn now to the arteries. The first branches of the aorta are a pair of small coronary arteries, which arise from the base of the aorta deep to the pulmonary trunk. The more conspicuous **left coronary artery** is accompanied by the large vein of the heart (the **great cardiac vein**, or *vena cordis magna*) and passes diagonally across the ventral surface of the heart between the right and left ventricles (Fig. 4-2A). You will see the right coronary artery when the heart is removed (Section E). Both coronary arteries carry blood to capillary beds in the heart wall.

To see the arteries, which lie dorsal to the veins in the arm and neck, you may use one of two approaches. You can try to dissect the arteries by working around the veins and pushing the veins to one side. This approach has the advantage of preserving the veins for later review, but it is very challenging. Alternatively, access the arteries by cutting through the left brachiocephalic and costo-cervical veins and pushing these veins aside. As you find the arteries, you may see accompanying veins that you may have previously missed.

The arch of the aorta gives off the **brachiocephalic trunk** and the **left subclavian artery** (Fig. 4-4). Trace the brachiocephalic trunk. It soon branches into a **right subclavian artery** and a pair of **common carotid arteries**. Several branches arise from each subclavian artery, but some are very deep and not easily seen. A deep **costocervical trunk** supplies the deep muscles of the neck. One of its branches, the **highest intercostal artery**, may be seen going to the cranial intercostal spaces. Another deep artery, the **vertebral artery**, runs cranially through canals in the transverse processes of the cervical vertebrae, enters the skull, and supplies parts of the brain.

The most superficial and easily seen branches of the subclavian artery are an **internal thoracic artery**, which you may have seen beside the corresponding vein, and a **thyrocervical trunk**, which supplies the thyroid gland and superficial parts of the neck. After crossing the first rib, the subclavian artery, now known as the **axillary artery**, gives off an **external thoracic artery**, which supplies the lateroventral parts of the thorax surface (Fig. 4-4). The axillary artery becomes the **brachial artery** as it enters the arm. The common carotid artery extends cranially with the internal jugular vein. At the base of the head, the common carotid artery divides into **external** and **internal carotid arteries**, which respectively supply the superficial and deeper parts of the head.

Dissect between the common carotid artery and the internal jugular vein. The white strand between them is the **vagus nerve** (Fig. 4-5), a part of the autonomic nervous system that supplies most of the viscera with parasympathetic nerve fibers (Exercise 7). A narrower **sympa-**

thetic cord, also a part of the autonomic nervous system, lies deep to the vagus and is closely bound to the vagus in much of the neck. Trace these nerves caudally (Fig. 4-5). They separate near the base of the neck. The vagus passes superficial (i.e., ventral) to the subclavian artery and extends to the esophagus, where you have seen it (Exercise 3, Section C). Do not confuse it with the **phrenic nerve**, which you saw earlier in this exercise and in Exercise 3, Section C. If you trace the phrenic nerve cranially from the thorax, it curves laterally at the level of the first rib and extends deeply toward the origin of cervical nerves. A small branch of the sympathetic cord follows the vagus nerve to the heart and lungs, but most of the sympathetic cord passes deep (dorsal) to the arteries you have dissected and onto the inside of the chest wall near the point of attachment of the ribs to the vertebral column. We will return to it later (Exercise 7).

After giving rise to the left subclavian artery, the aorta curves caudally and descends along the dorsal thoracic wall. Pull the left lung ventrally and follow the aorta. It soon receives the ductus arteriosus from the pulmonary trunk (Fig. 4-3) and gives rise to many pairs of **intercostal**

Veins and Arteries of the Neck, Head, and Thorax

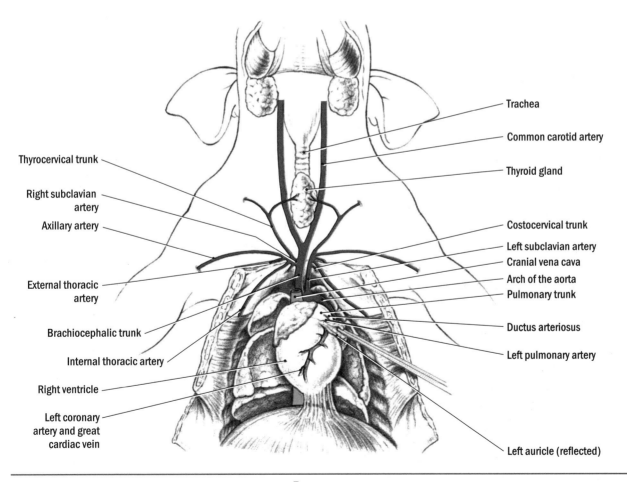

Thyrocervical trunk

Right subclavian artery

Axillary artery

External thoracic artery

Brachiocephalic trunk

Internal thoracic artery

Right ventricle

Left coronary artery and great cardiac vein

Trachea

Common carotid artery

Thyroid gland

Costocervical trunk

Left subclavian artery

Cranial vena cava

Arch of the aorta

Pulmonary trunk

Ductus arteriosus

Left pulmonary artery

Left auricle (reflected)

FIGURE 4-4
Ventral view of the heart and cranial arteries of a fetal pig after removal of the veins.

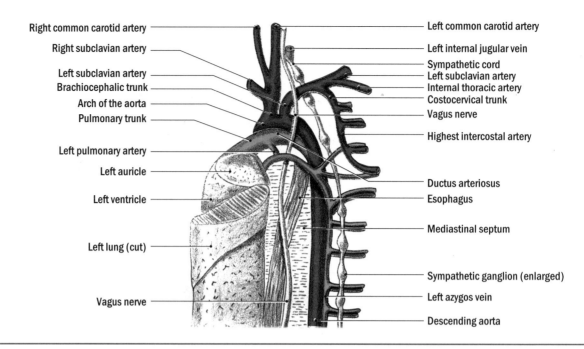

Right common carotid artery — Left common carotid artery
Right subclavian artery — Left internal jugular vein
Left subclavian artery — Sympathetic cord
Brachiocephalic trunk — Left subclavian artery
Arch of the aorta — Internal thoracic artery
Pulmonary trunk — Costocervical trunk
Left pulmonary artery — Vagus nerve
Left auricle — Highest intercostal artery
Left ventricle — Ductus arteriosus
Left lung (cut) — Esophagus
Vagus nerve — Mediastinal septum
Sympathetic ganglion (enlarged)
Left azygos vein
Descending aorta

FIGURE 4-5
Ventrolateral view of blood vessels and nerves at the base of the neck and dorsal thoracic wall of a fetal pig.

Veins and Arteries Caudal to the Diaphragm

arteries, which supply the intercostal muscles between the ribs caudal to those supplied by the **highest intercostal artery** (Fig. 4-5).

Intercostal veins drain the intercostal spaces. Most of those from both the left and the right side of the body collect to form a **left azygos vein**, which turns ventrally across the left pulmonary vein (Fig 4-5) and then crosses the dorsal side of the heart, where it forms the **coronary sinus** (Fig. 4-8). The coronary sinus receives several small **cardiac veins** from the heart wall and then enters the right atrium beside the entrance of the caudal vena cava.

The cranial arteries and veins are very similar in human beings and pigs. Major differences are the origin of the left common carotid artery directly from the arch of the aorta and the presence of both left and right azygos veins in the human being.

D. VEINS AND ARTERIES CAUDAL TO THE DIAPHRAGM

D.1. Blood Vessels of the Digestive Organs

The abdominal and pelvic viscera, hind legs, and tail are supplied by branches of the aorta and are drained ultimately by the caudal vena cava. The digestive organs and the spleen, however, are drained first by a group of veins that carries blood to the liver. These veins constitute the **hepatic portal system**. A portal system is defined as veins that lead from capillaries in one organ to capillaries in another organ rather than directly to the heart.

Blood from organs drained by the hepatic portal system passes through the hepatic sinusoids within the liver. Because these sinusoids lack the complete lining of endothelial cells that is present in true capillaries, blood comes into direct contact with hepatic cells.

Many metabolic conversions occur in the liver and can be illustrated by several examples. Just after a meal, surplus absorbed sugar (glucose) is stored in the liver, largely in the form of glycogen. Between meals, sugar may be released by the liver so that the glucose content of the blood is kept within narrow limits. Amino acids are deaminated in the liver, and their amino groups form ammonium ions and ammonia, which are converted into urea, later to be released from the body by the kidneys. Some absorbed toxins are removed from the blood or broken down (e.g., alcohol). Certain products from the breakdown of senescent red blood cells in the spleen are salvaged; others are removed and excreted as bile pigments.

Break the greater omentum, push the stomach and spleen cranially and the intestines caudally. Separate the colon from the tail of the pancreas, if this has not been done earlier, and carefully free the tail of the pancreas and push it aside, or dissect it away. Often the hepatic portal system is injected through the umbilical vein. If not, the hepatic portal system probably contains blood and can be seen deep to the tail of the pancreas (Fig. 4-6).

The veins are accompanied by small arteries. The **left gastric vein** drains the stomach near its lesser curvature. The **left gastroepiploic vein** drains the stomach along its greater curvature and joins the **lienic vein**, which drains the spleen (lien). The lienic and left gastric veins unite to form the **lienogastric vein**. This vein joins the **cranial mesenteric vein** to form the **hepatic portal vein**. As the hepatic portal vein passes forward dorsal to the pylorus, it receives a small **gastroduodenal vein** which drains the duodenum, pylorus, and pancreas. The hepatic portal vein continues to the liver in the lesser omentum. As it enters the liver, it is joined by branches of the **umbilical vein** returning blood from the placenta.

Cut through the left side of the diaphragm, push the abdominal viscera to the specimen's right, and trace the **aorta** into the abdominal cavity. It passes caudally between the two **kidneys**, accompanied by the **caudal vena cava** (Fig. 4-7).

Notice that the aorta, caudal vena cava, and kidneys lie against the back muscles deep to the parietal peritoneum. This position is described as **retroperitoneal**. You must carefully peel off the peritoneum to see this area clearly.

As you peel off the peritoneum between the left kidney and the aorta, look for a light-colored band of tissue adjacent to the cranial end of the kidney. It often adheres to the peritoneum. This is the suprarenal gland, or **adrenal gland**. The medulla of this endocrine gland secretes epinephrine, or adrenalin, a hormone that helps the sympathetic nervous system adjust the body to stress (the flight or fight reaction, see Exercise 7). Its cortex secretes mineralocorticoid hormones, which regulate salt and mineral metabolism, and glucocorticoid hormones, which have various functions in protein and carbohydrate metabolism as well in counteracting inflammations. An androgen hormone promotes protein synthesis and growth and resembles the male sex hormone, testosterone. At abnormally high levels in females, this hormone may have a masculinizing effect.

Carefully clear the surface of the aorta between the diaphragm and the kidney. You may have to remove the left adrenal gland. As you do so, you may notice one or more strong, white, fibrous strands extending from the sympathetic cord on the dorsal thoracic wall (Fig. 4-5). These are **splanchnic nerves**, which go to a network of

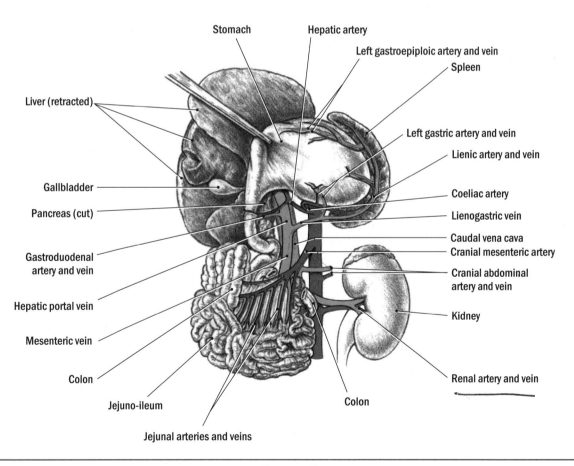

Veins and Arteries Caudal to the Diaphragm

FIGURE 4-6
Ventral view of the arteries and veins supplying the digestive organs of a fetal pig. The stomach and spleen have been pulled cranially, the intestines have been pulled caudally, and the tail of the pancreas has been removed.

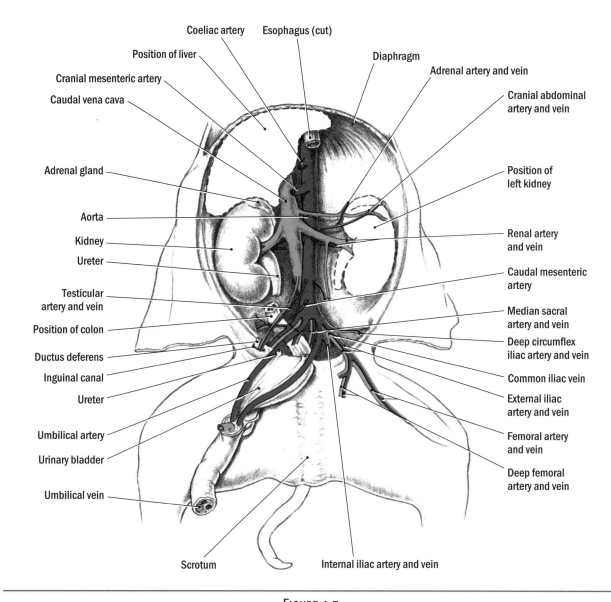

Coeliac artery   Esophagus (cut)
Position of liver
Cranial mesenteric artery
Caudal vena cava
Diaphragm
Adrenal artery and vein
Cranial abdominal artery and vein
Adrenal gland
Position of left kidney
Aorta
Kidney
Ureter
Renal artery and vein
Caudal mesenteric artery
Testicular artery and vein
Position of colon
Ductus deferens
Inguinal canal
Ureter
Median sacral artery and vein
Deep circumflex iliac artery and vein
Common iliac vein
External iliac artery and vein
Umbilical artery
Urinary bladder
Femoral artery and vein
Deep femoral artery and vein
Umbilical vein
Scrotum   Internal iliac artery and vein

Veins and Arteries Caudal to the Diaphragm

FIGURE 4-7
Ventral view of arteries and veins lying against the dorsal abdominal and pelvic walls of a fetal pig.
The drawing has been made as if the digestive organs, ureters, and left kidney had been removed,
but these organs should only be pushed aside during the dissection.

tough nerve fibers and masses of nerve cell bodies (**collateral sympathetic ganglia**), which cover part of the aorta and its branches. We will discuss these in Exercise 7, but you must pick them up or push them away carefully to see the blood vessels.

Return to the aorta where it perforates the diaphragm. You may notice one or more small **phrenic arteries** that supply the diaphragm. The phrenic arteries may arise from the aorta as it passes through the diaphragm or from the first branch of the abdominal segment of the aorta, the median **coeliac artery** (Fig. 4-7). The coeliac artery soon divides into two branches (Fig. 4-6): the lienic artery soon gives off a **left gastric artery**, which supplies

the stomach near its lesser curvature. After supplying the spleen, the lienic artery continues as the **left gastroepiploic artery** along the greater curvature of the stomach. The hepatic artery, after giving off branches to parts of the pancreas, duodenum, and stomach, continues to the liver, accompanied by the hepatic portal vein. A small **gastroduodenal artery** arises from the hepatic artery to supply the duodenum, pylorus, and pancreas. The liver receives large amounts of blood through the hepatic portal system, but, in an adult, the hepatic artery delivers oxygen-rich blood to this metabolically very active organ. All of these vessels follow veins of the hepatic portal system previously seen.

The next branch of the abdominal portion of the aorta is the median **cranial mesenteric artery** (Fig. 4-7). It gives rise to numerous **jejunal arteries** that supply most parts of the small intestine and part of the large intestine.

Find the caudal vena cava just caudal to the liver (Fig. 4-7) and trace it cranially through the liver by dissecting away tissue from the right dorsal surface of the caudate lobe of the liver. Although the caudal vena cava extends through the liver, it does not carry blood to the liver; rather, it receives blood from the liver by several **hepatic veins**, which drain the hepatic sinusoids. In the fetus, some branches of the umbilical vein lead to liver sinusoids, but much of the umbilical blood bypasses the sinusoids in a special channel and enters the caudal vena cava directly (see later). You may see one or more **phrenic veins**, which drain the diaphragm and enter the vena cava as it goes through the diaphragm into the thorax.

D.2. Blood Vessels on the Dorsal Abdominal and Pelvic Walls

Return to the aorta and caudal vena cava at the level of the kidneys. Caudal to the cranial mesenteric artery, the aorta gives rise to a pair of large **renal arteries**, which supply the kidneys. Paired **renal veins** accompany the arteries and enter the caudal vena cava.

Push the intestine to the right side of the specimen, but do not remove it. Lift up the lateral edge of the left kidney and dissect deep to it. You will soon see the **cranial abdominal artery** and **vein** that supply the back in this region and also give rise to the small **adrenal artery** and **vein**. The cranial abdominal arteries arise from the aorta. The cranial abdominal veins enter the caudal vena cava or the renal veins.

Notice the **urinary bladder**, which lies against the midventral strip of the body wall previously reflected, and find the reproductive organs. If your specimen is a male, the **testes** will have partially descended into a skin sac, the scrotum, through a passage in the body wall known as the **inguinal canal** (Fig. 4-7). Do not dissect or damage the scrotum and inguinal canal until you study the urogenital system. If your specimen is a female, the paired, coiled **uterine horns** and the **ovaries** can be seen lying dorsal to the base of the urinary bladder. Small testicular or ovarian blood vessels supply the gonads. The paired **testicular** or **ovarian arteries** arise from the aorta caudal to the renal arteries; the **testicular** or **ovarian veins** enter the caudal vena cava, or the left vein may enter the left renal vein.

A median **caudal mesenteric artery** leaves the aorta near the gonadal arteries and soon branches. Its cranial branch extends cranially along the colon and supplies much of the colon.

Carefully dissect away connective tissue from the dorsal body wall lateral to the caudal section of the aorta. Also free, but do not destroy, the large convoluted duct (the **ureter**) passing from the kidney to the base of the urinary bladder. If you lift the aorta slightly, you may notice several pairs of **lumbar arteries** arising from its dorsal surface and going to deep back muscles. **Lumbar veins** accompany the arteries and enter the caudal vena cava.

Dorsal to the ureter, the aorta gives rise to a pair of **external iliac arteries** (Fig. 4-7). Trace one. A lateral branch, the **deep circumflex iliac artery**, soon leaves to supply some of the pelvic muscles. It is accompanied by the **deep circumflex iliac vein**. The external iliac artery, now accompanied by the **external iliac vein**, continues caudally. After perforating the muscle layers of the body wall, the blood vessels enter the leg and are now called the

Veins and Arteries Caudal to the Diaphragm

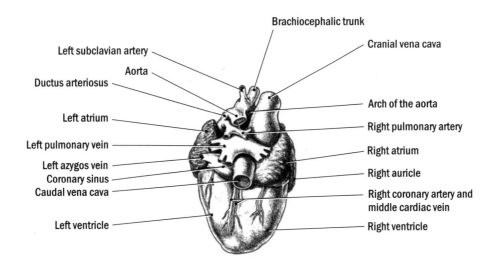

FIGURE 4-8
Dorsal view of the heart and its great vessels of a fetal pig.

femoral **artery** and **vein**. You must separate the muscle layers of the abdominal wall from the skin over the caudal abdominal wall and thigh to see these vessels clearly. A **deep femoral artery** and **vein** leave the medial side of the proximal end of the femoral blood vessels.

Just caudal to the external iliac arteries, the aorta appears to bifurcate, forming two large blood vessels that extend onto the ventral body wall along the sides of the urinary bladder and continue into the umbilical cord. The distal parts of these blood vessels are the **umbilical arteries**, which carry blood from the fetus to the placenta. The short proximal part of each of these vessels, together with a small dorsal branch extending into the pelvic cavity, is the **internal iliac artery**. In the adult, this dorsal branch and the base of the internal iliac artery are of equal size, because the umbilical artery atrophies except for a small part supplying the urinary bladder.

The proximal part of the internal iliac artery is accompanied by an **internal iliac vein**. Internal iliac and external iliac veins unite to form a short **common iliac vein**, which joins the caudal vena cava. The terminal branch of the aorta is a small **median sacral artery** that arises deep to the bifurcation of the internal iliac arteries and extends caudally along the dorsal pelvic wall. It enters the tail as the **caudal artery**. It is accompanied by a **median sacral** and **caudal vein**, which usually enters one of the common iliac veins. These will be seen more clearly while studying the urogenital organs (Exercise 5).

The pattern of these vessels is similar in human beings, except that the external and internal iliac arteries arise from a common iliac artery rather than independently from the aorta.

Root of the
Lung and
Structure
of the Heart

E. ROOT OF THE LUNG AND STRUCTURE OF THE HEART

Carefully cut through the aorta and caudal and cranial venae cavae respectively leaving and entering the heart. Leave rather long stumps of the vessels attached to the heart; in particular, cut the aorta distal to the entrance of the ductus arteriosus. Lift up the cranial end, or base, of the heart, dissect beneath it, and gradually pull the heart and attached vessels caudally. The **pulmonary arteries** can now be seen more clearly and traced to the lungs. **Pulmonary veins**, which return blood from the lungs to the left atrium, can be seen caudal to the distal part of the pulmonary arteries.

Find the trachea at the base of the neck and trace it to the lungs. You can now see the **bronchi** branching from the trachea and entering the lungs (see Exercise 3). One bronchus arises from the right side of the trachea and passes to the cranial lobe of the right lung. Further caudally, the trachea bifurcates into two large bronchi, one leading to the rest of the right lung, the other to the left

66

lung. The area where the pulmonary blood vessels and bronchi enter and leave the lung is called the **root of the lung**. Cut through the pulmonary blood vessels near the lungs and lift the heart from the thorax.

The heart creates the hydrostatic pressure necessary to keep the blood flowing. Its basic structure can be seen quite well in the heart of a fetal pig, but demonstration dissections of an adult sheep or beef heart should also be examined.

Orient the heart and reidentify the blood vessels that enter and leave it (Figs. 4-2 and 4-3). Pick away connective tissue to get a clearer view of its dorsal surface (Fig. 4-8). The bifurcation of the pulmonary trunk into the **pulmonary arteries** can now be seen clearly. Just caudal to them, the **pulmonary veins** converge to enter the left atrium. Caudal to the pulmonary veins, the **left azygos vein** enters the **coronary sinus**, which crosses the dorsal side of the heart to enter the right atrium. The **right coronary artery**, which you have identified earlier, and the **middle cardiac vein** (*vena cordis media*) course between the right atrium and ventricle and then turn caudally between the left and right ventricles.

Open the heart by making an incision through the lateral wall of each atrium and its auricle (incisions 1 and 2, Fig. 4-9) and carefully remove the injection mass or coagulated blood that it contains. Observe the thinness of the atrial walls. The atria simply collect blood from the veins during ventricular contraction, or systole. During ventricular relaxation, or diastole, much of the blood is sucked into the ventricles, but their final filling is aided by the contraction of the atria. Notice the entrances of the cranial and caudal venae cavae and the coronary sinus, which drain the body and heart muscle itself, into the right atrium, and the entrances of the pulmonary veins, which drain the lungs, into the left atrium. At the caudal end of each atrium, there is an opening into a ventricle guarded by a valve. The valves will be seen more clearly as the dissection progresses.

In an adult, right and left atria are separated from each other by the **interatrial septum**, but in the fetus, they are partly connected by the **foramen ovale**, which perforates the interatrial septum. A valve in the foramen ovale permits blood to pass only from the right to the left atrium and, thereby, to bypass the lungs. The entrance to the foramen ovale lies near the dorsal wall of the heart, just cranial to the atrial entrance of the caudal vena cava into the right atrium (Fig. 4-10).

Extend the incision that you made through the right atrial wall into the caudal vena cava to see the foramen ovale clearly. The foramen ovale permanently closes at birth but leaves a depression in this region, the **fossa ovalis**, which can be seen in an adult heart. To the left of the foramen ovale and cranial to the opening between the right atrium and ventricle, you will also find the entrance of the coronary sinus.

A specialized bulge of cardiac muscle (the **sinuatrial node**, or pacemaker) lies in the atrial wall between the

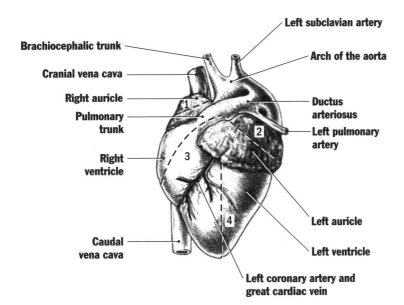

FIGURE 4-9
Ventral view of the heart of a fetal pig, showing the location
of the incisions to be made to open the heart.

foramen ovale and the entrance of the cranial vena cava, but it may be difficult to identify. The activity of the pacemaker initiates the rhythmic contraction of cardiac muscle (Exercise 2).

Again notice the coronary arteries on the dorsal and ventral surfaces of the heart, demarcating the right from the left ventricle. Cut open the right ventricle by making a diagonal incision through its ventral wall and extending the incision into the pulmonary trunk (incision 3, Fig. 4-9). Carefully clean out the inside of the ventricle and pulmonary trunk. Notice the thick muscular wall of the ventricle.

The three flaps of the **right atrioventricular,** or tricuspid, **valve** may be seen protruding into the ventricle

Root of the Lung and Structure of the Heart

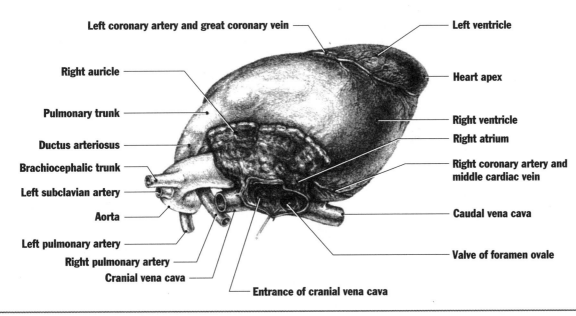

FIGURE 4-10
View of the right side of the heart of a fetal pig, with the right atrium cut open to show the foramen ovale.

from the right atrioventricular opening, but they are often torn when the injection mass is removed. If so, observe them on a demonstration dissection of a heart. The margins of the valves are anchored to the ventricular wall by delicate **tendinous cords**. These permit the flaps to close against one another, yet prevent them from everting into the atrium when the ventricle contracts. Look into the proximal end of the pulmonary trunk and notice the three semilunar-shaped pockets of the **pulmonary valve**. This valve prevents blood in the pulmonary trunk from backing up into the right ventricle during ventricular relaxation. Casts of these pockets can often be seen on the injection mass removed from the pulmonary trunk.

Open the left ventricle by making a longitudinal incision through its ventral wall and extending the incision toward the origin of the aorta (incision 4, Fig. 4-9). Clean it out. The muscular wall of the left ventricle is thicker than that of the right ventricle. This correlates with the different peripheral resistance to be overcome by each ventricle: to all of the body by the left ventricle, only to the lungs by the right ventricle. The **left atrioventricular**, or bicuspid, **valve** is similar to the right one but has only two flaps. The **aortic valve**, which resembles the pulmonary valve, lies in the base of the aorta.

F. FETAL AND NEONATAL CIRCULATION

F.1. Fetal Circulation

The placenta—rather than the digestive tract, lungs, and kidneys—provides the embryo with nutrients and oxygen and removes carbon dioxide and nitrogenous wastes from its body. Accordingly, the pattern of fetal circulation differs in several important aspects from the circulation of an adult (Fig. 4-11). Using the following account of fetal circulation, trace the blood vessels and identify the parts of the heart on your own specimen.

Blood rich in oxygen and nutrients and low in waste products enters the embryo by way of the umbilical vein. Some of this blood joins the hepatic portal system and flows through liver sinusoids, but much of it passes directly through the liver, bypassing its sinusoids, in a channel known as the **ductus venosus**, which enters the caudal vena cava. Thus, the oxygen-rich blood coming from the umbilical vein through the ductus venosus mixes with the oxygen-depleted blood coming from the body through the caudal vena cava, as well as with the oxygen-depleted blood coming from the liver sinusoids through hepatic veins. Because of the close anatomical relationship between the entrance of the caudal vena cava into the right atrium and the foramen ovale, most of the blood coming from the caudal vena cava flows directly through the foramen ovale into the left atrium and mixes very

little with other blood entering the right atrium. From the left atrium, this blood from the caudal vena cava enters the left ventricle and continues on into the aorta, by which it is distributed to the body. Notice that the heart muscle, head, and front legs receive "mixed" blood from the aorta before oxygen-depleted blood from the right ventricle enters the aorta through the ductus arteriosus.

The blood returning from the heart muscles, head, and front legs enters the right atrium and most of it continues into the right ventricle, because the entrances of the cranial vena cava and coronary sinus are directed toward the right atrioventricular opening. This oxygen-depleted blood leaves the heart through the pulmonary trunk toward the lungs.

Throughout much of fetal life, the lungs are not sufficiently developed to receive all the blood coming from the right ventricle. Moreover, because the lungs are not expanded by air, they are compact organs that offer a relatively great resistance to blood flow. Consequently, only a small volume of blood flows through them and returns to the left atrium in the pulmonary veins.

Most of the blood in the right ventricle bypasses the lungs, traveling through the ductus arteriosus to the aorta, where it mixes with the blood coming from the left ventricle. This mixed blood is distributed by the aorta to the rest of the body and through the umbilical arteries to the placenta.

Within the placenta, fetal and maternal blood come so close together that an exchange of nutrients, oxygen, and waste products occurs primarily by diffusion. Normally, there is no mixing of the two kinds of blood.

The foramen ovale and ductus arteriosus not only carry a considerable volume of blood bypassing the lungs but also have another important role. To ensure the normal development of their thick, muscular walls, the ventricles must pump a reasonable volume of blood even though little is being pumped to the lungs and little returns from the lungs to be pumped to the body. The ductus arteriosus permits the right ventricle to pump a volume of blood greater than that needed by the developing lungs. For this reason, the ductus arteriosus is sometimes called the "exercise channel" of the right ventricle. Similarly, the foramen ovale supplies the left side of the heart with the volume of blood necessary for its development even though little blood returns from the lungs. The foramen ovale is the "exercise channel" of the left ventricle.

F.2. Changes at Birth

In late fetal life, the foramen ovale becomes smaller relative to the rest of the heart, and the lumen of the ductus arteriosus becomes relatively narrower. These two changes cause an increasing amount of blood to flow through the lungs. But the adult circulatory pattern is not assumed until after birth. At that time, the lungs fill with air, the

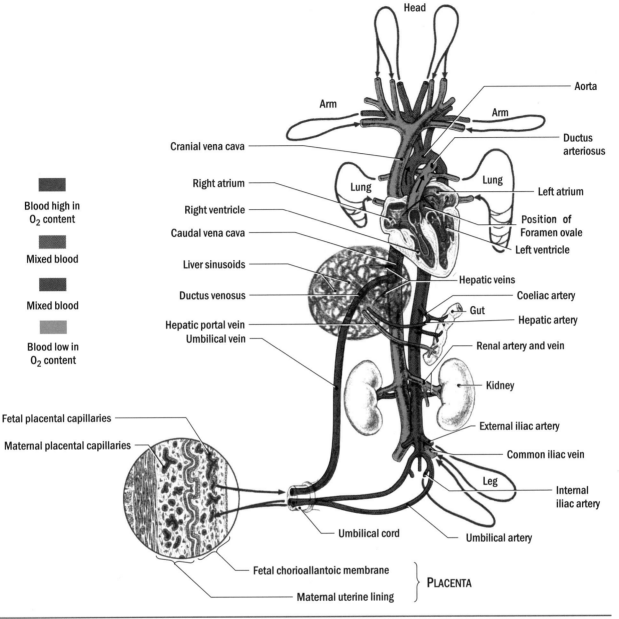

Blood high in O₂ content

Mixed blood

Mixed blood

Blood low in O₂ content

Fetal and Neonatal Circulation

FIGURE 4-11
Diagrammatic ventral view of the course of blood flow in a fetal pig. *Inset*: Microscopic view of a part of the placenta.

lung capillaries are able to expand, and the resistance to blood flow through the lungs is reduced. As more blood coming from the right ventricle passes to the lungs, more blood returns from the lungs to the left atrium and ventricle. The increased blood volume and pressure in the left atrium keep the flaps of the valve in the foramen ovale closed, so blood entering the right atrium is no longer shunted to the left atrium.

The ductus arteriosus remains open for a few days after birth, but blood flows through it in a direction that is the reverse of that in a fetus. This reverse flow occurs because the resistance in the pulmonary circuit is now less than the resistance in the systemic circuit to the body. As a consequence, some of the oxygen-rich blood leaving the heart through the aorta enters the pulmonary arteries through the ductus arteriosus and passes through the lungs a second time. This double passage of a fraction of the blood through the lungs constitutes the **neonatal circulation**.

Blood is more thoroughly aerated at this critical time immediately after birth when fetal hemoglobin is being replaced by the adult type of hemoglobin. Fetal hemoglobin

69

evolved in placental mammals as one of many adaptations for intrauterine life. Fetal hemoglobin has a higher affinity for oxygen than adult hemoglobin does, and this enables fetal hemoglobin in the placenta to take oxygen away from the maternal blood. However, fetal hemoglobin with its high affinity for oxygen would not discharge oxygen efficiently in the tissues of a more active adult. The adult type of hemoglobin, although it has a lower affinity for oxygen, does become nearly fully saturated with oxygen in the lungs, which are now filled with air that has a higher oxygen content than maternal blood. Its lower affinity for oxygen enables it to give up oxygen easily in the tissues.

The neonatal circulatory pattern lasts only from a few hours to a day or so. The ductus arteriosus then contracts and is filled in with connective tissue, and the adult circulatory pattern is established. A permanently occluded remnant of the ductus arteriosus remains as the **ligamentum arteriosum** (Fig. 4-1). The flaps of the valve of the fetal foramen ovale fuse. The site of the foramen ovale is represented in the adult by a depression known as the **fossa ovalis**, as mentioned earlier. The umbilical vein and ductus venosus atrophy, as do the parts of the umbilical arteries distal to the urinary bladder.

In a transformation as complex as this, it is not surprising that anomalies sometimes occur. The failure of the ductus arteriosus or foramen ovale to close completely, for example, results in insufficient oxygen uptake by the blood—a human infant suffering from this condition is known as a "blue baby."

G. BLOOD

The blood vessels that you have been dissecting carry blood between different parts of the body. They are important pathways, but it is the blood that performs the transport, defense, and homeostatic functions of the circulatory system. Blood consists of a liquid **plasma** and several kinds of blood cells carried in the plasma. If blood smears from human beings or other mammals are available, or if you have the materials and stain needed to prepare them, the major types of blood cells can be examined.

G.1. Erythrocytes

Red blood cells, or **erythrocytes**, are the most abundant type of blood cells, numbering 4.1 to 6.0 million per microliter of blood in an adult man and 3.9 to 5.5 million per microliter of blood in an adult woman. They appear as pinkish, circular cells about 8 micrometers in diameter (Fig. 4-12). Their nuclei, mitochondria, and most other cell organelles are lost in the course of their maturation in mammals, so they are shaped like biconcave disks. This is

most evident in an edge view; but even in a surface view, the fact that the center of the cell is thinner than the periphery is usually discernible. This shape gives the cell an even larger surface area relative to its volume than if it were a sphere.

Erythrocytes are filled with the respiratory pigment **hemoglobin**, which binds reversibly with oxygen, taking oxygen up in the lungs and releasing most of it in the tissues. Most of the carbon dioxide released by the tissues is carried in the plasma as carbonic acid or its ions and salts, but some also combines with hemoglobin and is transported to the lungs for release.

Because erythrocytes lack nuclei and most cell organelles, they live only a short time, about 100 to 120 days. Cells lining the spleen and liver sinusoids capture and digest (**phagocytose**) senescent red blood cells. Phagocytosed cells are replaced by new ones that are continuously produced through mitosis from nucleated stem cells in the red bone marrow.

G.2. Leukocytes

The other blood cells are in life colorless, or white, and are referred to collectively as **leukocytes**. There are several kinds of leukocytes, which number in aggregate about 5,000 to 9,000 per microliter of human blood. All leukocytes are nucleated, and most are larger than the erythrocytes (Fig. 4-12). Some have diameters as large as 20 micrometers. For the purpose of description, it is convenient to divide them into two groups: those having many conspicuous cytoplasmic granules (**granular leukocytes**) and those having very few cytoplasmic granules or none at all (**agranular leukocytes**). The granules have distinctive colors when treated with Wright's stain.

All leukocytes are capable of amoeboid movement or, in other words, of squeezing between the endothelial cells of capillary walls and moving around within the tissues. They aggregate in areas of infection and inflammation. The relative numbers of the different types of leukocytes present in the blood are frequently diagnostic of certain diseases.

Lymphocytes are the most common agranular leukocytes, comprising from 20 to 30 percent of the leukocyte population. Most are slightly larger than erythrocytes and have a large, nearly spherical nucleus surrounded by only a thin layer of cytoplasm. Circulating lymphocytes contain few cytoplasmic organelles and are in a "resting" state, but they can change into active cells that play a critical role in the body's defense and in the development of immunity. B-lymphocytes can change into cells that produce antigens against invading antibodies. T-lymphocytes attack invading antigens directly in a cell-mediated reaction.

Monocytes are much larger and less common agranular leukocytes that make up from 2 to 8 percent of the leukocytes. The nucleus of a monocyte is typically kidney shaped, and its cytoplasm is abundant. Monocytes

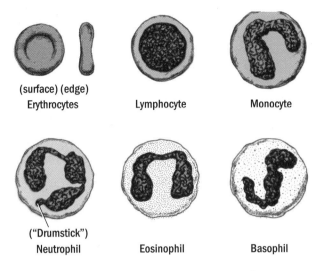

(surface) (edge)
Erythrocytes　　Lymphocyte　　Monocyte

("Drumstick")
Neutrophil　　Eosinophil　　Basophil

FIGURE 4-12
The major types of mammalian blood cells can
be distinguished by cell size and shape, nuclear shape,
and cytoplasmic granulation.

change into tissue **macrophages**. These large and versatile cells play a role in the maintenance of normal tissues, but they are also important in phagocytosis, the development of immunity, and the repair of injured tissues. Monocytes are essentially macrophages that are in transit from the bone marrow, where they are produced, to the tissues.

Neutrophils, which constitute from 60 to 70 percent of all leukocytes, are the most common granular leukocytes. The nucleus is elongated, twisted, and usually constricted into three lobes, which are connected by thin chromatin threads. Occasionally, part of the nucleus of a neutrophil in a female individual is set off as a small appendage called a "**drumstick**," which consists of the chromatin of the female sex chromosomes. The cytoplasm has a fine, barely perceptible granulation that stains a light purple. Neutrophils can phagocytose microbes within the bloodstream, and they also collect in large numbers at sites of injury or infection, phagocytosing bacteria and dead cells. Their granules release enzymes that digest ingested particles. Therefore, the granules decrease in number as

ingested particles are broken down. Neutrophils are one of the body's first lines of defense.

Eosinophils make up from 2 to 5 percent of the leukocytes. The nucleus usually consists of two lobes connected by a chromatin thread. Many relatively large, reddish granules fill the cytoplasm. Products released by the granules can kill some parasitic larvae within the circulatory system. Eosinophils collect in connective tissues near a body surface, such as the dermis and the lining of the digestive and respiratory systems. Within the tissues, they appear to phagocytose various inflammatory substances and help rid the body of inactive antigen–antibody complexes.

Basophils are the rarest of the leukocytes, comprising only 0.5 to 2.0 percent of the leukocyte population. The elongated and twisted nucleus frequently has an S-like configuration. The cytoplasm contains bluish granules of various sizes. The granules of basophils contain histamine and heparin, which are released into injured tissues. Histamine dilates capillaries and increases their permeability to other leukocytes. Heparin is an anticoagulant that keeps blood flowing to injured tissues.

Some lymphocytes live for years, thereby providing the basis for immunological memory and the development of immunity. But most leukocytes live for only a few days, because they are continuously being lost through the mucous membranes of the digestive, respiratory, and urinary tracts. Lymphocytes multiply and are stored in lymph nodes and other lymphoid tissues. All other leukocytes are produced through mitosis from nucleated stem cells in the red bone marrow.

Blood

G.3. Blood Platelets

Look for small groups of granules surrounded by bits of cytoplasm that are scattered among the other cells. These are the **blood platelets**, which are simply small blobs of cytoplasm that have budded off certain giant cells (**megakaryocytes**) in the bone marrow. They help protect the body from excess blood loss at the site of an injury: first, by clumping and helping to plug an injured blood vessel and, second, by releasing phospholipids. Phospholipids from injured platelets and **thromboplastin** from injured tissues initiate a series of enzyme-mediated reactions that result in the transformation of the soluble plasma protein fibrinogen into fibrin threads that form a blood clot.

Urogenital System

T HE UROGENITAL SYSTEM consists of two systems, the excretory system and the genital system. The two systems serve very different functions but are structurally so closely associated with each other that they share certain ducts and passages. For this reason, it is convenient to study the parts of the two systems together. Their close anatomical association results from their development from adjacent embryonic precursors.

A. EXCRETORY SYSTEM

The excretory system consists mainly of the kidneys, which have a vital role in eliminating nitrogenous waste products of the protein metabolism by body cells. The kidneys also help maintain homeostasis by regulating the pH as well as the balance of water, salts, ions, sugars, and many other substances in the body fluids.

A.1. Dissection

The pair of **kidneys** (*renes*) has been seen, and probably partly uncovered, in previous dissections (Fig. 5-1; see also Exercises 3 and 4). The kidneys lie against the ventral surface of the back muscles and bulge into the peritoneal cavity. The serosa, or parietal peritoneum, covers only their ventral surface; hence, the position of the kidneys is described as **retroperitoneal** ("behind the peritoneum"). The narrow, bandlike **adrenal**, or suprarenal, **gland**, which lies adjacent to the craniomedial border of each kidney, has already been seen (Exercise 4). Try to save it as you carefully peel off the peritoneum and the capsule of connective tissue from the ventral surface of the pig's left kidney. In an adult, the kidneys are protected by a capsule of fat, called structural fat.

Each kidney is drained by a **ureter**, which leaves from the slightly indented medial margin of the kidney; the ureter is accompanied along its course by the **renal artery** and **vein**. Trace the ureter caudally, along the muscles of the back. Notice that it turns ventrally near the brim of the pelvis, passes deep to the ductus deferens (sperm duct) in a male, and enters the **urinary bladder**. The urinary bladder is attached by a ventral mesentery, the **ventral ligament of the bladder**, to the midventral strip of the abdominal wall, which you have previously turned back.

Trace the urinary bladder into the umbilical cord and notice that it continues as the **allantoic stalk**. The urinary bladder develops embryonically from the intraabdominal part of the allantois. Most of the allantois extends peripherally to the body as a large extraembryonic sac that helps to form the placenta (see Fig. 4-11 and Fig. 5-8).

The caudal end of the urinary bladder narrows to form a duct, the **urethra**, which dips into the pelvic cavity. Its subsequent course to the body surface will be seen during the dissection of the reproductive organs in Section B.

Return to one of the two kidneys and section it in the frontal plane of the body; that is, cut off its ventral half along the plane of the ureter (Fig. 5-1). A single-edged razor blade may be more effective in making this cut than a scalpel. Dissect away blood vessels in the half of the kidney left in the body, and also open the ureter with a fine pair of scissors. Notice that the ureter expands within the kidney to form a large chamber, the **renal pelvis**. The renal pelvis is partitioned into many smaller compartments, **renal calyces**, each of which receives a dark tuft of kidney

tissue. Each of these tufts is a **renal pyramid**. The renal pyramids collectively form the **renal medulla**. The peripheral, lighter-colored part of the kidney is the **renal cortex**.

A.2. Microscopic Structure of the Kidney

Examine a histological slide preparation of the kidney. The structural and functional unit of the kidney is the **nephron**. It is composed of the **glomerulus**, a capillary tuft, and the **renal tubule**, which receives fluid from the glomerulus and processes it to form urine. The various parts of a nephron have specific functions and are located in specific places within the cortex and medulla of the kidney (Figs. 5-2 and 5-3).

Blood coming from the **renal artery** (see Exercise 4) is distributed to **afferent arterioles** within the kidney. Blood in each afferent arteriole enters the capillaries of the glomerulus and then filters into the **glomerular capsule** (of Bowman). Together the glomerulus and the glomerular capsule form a structural and functional unit called the **renal corpuscle**.

The glomerular filtrate within the glomerular capsule is very similar in composition to blood, except that it does not contain blood cells and larger protein molecules, which are too large to pass through the exceptionally thin walls of the capillaries and glomerular capsule. The glomerular filtrate continues through the **proximal convoluted tubule** in the renal cortex, through the **loop of Henle** into and out of the renal medulla, and through the **distal convoluted tubule** back in the renal cortex. The distal convoluted tubules of several nephrons enter a **collecting tubule**, which extends from the renal cortex through the renal medulla to the tip of a renal pyramid. As the glomerular filtrate passes through the various parts of the renal tubule, water, ions, nutrients, and other substances needed by the body are absorbed by the walls of the renal tubule, in some cases actively and in some cases passively, and are returned to the bloodstream through the **peritubular capillaries**. Some water also may be absorbed by the collecting tubules. The peritubular capillaries are supplied by **efferent arterioles**, which collect blood from the glomerulus, and are drained by **renal venules**. The renal venules, in turn, lead to the **renal veins**. Waste products—in particular, ammonia and urea—remain in the glomerular filtrate. In this manner, the metabolic waste products become more and more concentrated as the glomerular filtrate passes through renal and collecting tubules, finally becoming urine by the time it leaves the renal pyramid to enter the ureter.

B. REPRODUCTIVE SYSTEM

The reproductive system produces and transports the gametes for sexual reproduction and, in female mammals, serves also to protect and nourish the embryo until it is

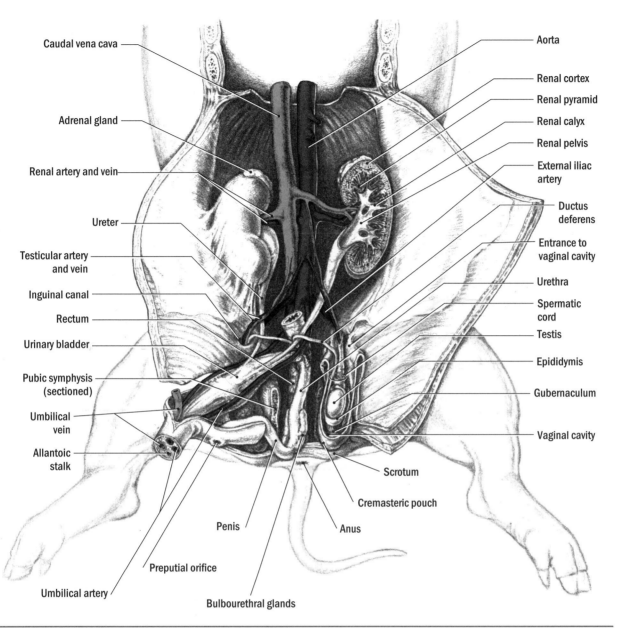

Caudal vena cava

Adrenal gland

Renal artery and vein

Ureter

Testicular artery
and vein

Inguinal canal

Rectum

Urinary bladder

Pubic symphysis
(sectioned)

Umbilical
vein

Allantoic
stalk

Penis

Preputial orifice

Umbilical artery

Bulbourethral glands

Aorta

Renal cortex

Renal pyramid

Renal calyx

Renal pelvis

External iliac
artery

Ductus
deferens

Entrance to
vaginal cavity

Urethra

Spermatic
cord

Testis

Epididymis

Gubernaculum

Vaginal cavity

Scrotum

Cremasteric pouch

Anus

*Reproductive
System*

FIGURE 5-1
Ventral view of the urogenital system of a male fetal pig.
The pelvic canal has been cut open, and the left kidney has been sectioned
in the frontal plane of the body. The left cremasteric pouch also has been cut open.
The liver and most of the digestive tract have been removed for the sake of clarity.

developed enough to be born. Parts of the testes and ovaries also function as endocrine glands that produce hormones which regulate many aspects of growth and reproduction. Although you will dissect the reproductive system of only one sex, you should use another student's specimen to study the opposite sex. In other words, you should be particularly careful in dissecting your own specimen—as careful as if you were preparing a demonstration preparation, which, in effect, you are.

B.1. Male Reproductive Organs

Notice again the external pouch, the **scrotum**, part of which can be seen ventral to the anus (see Exercise 1, Fig. 1-2). The scrotum is essentially a skin pouch that contains paired extensions of the abdominal wall and abdominal cavity in which the testes are housed. You will expose these pouches soon. At an early stage of embryonic development, the testes are located retroperitoneally just caudal to the kidneys; but during subsequent development, they

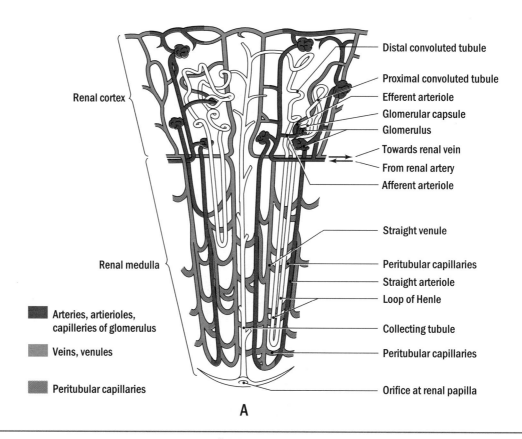

Renal cortex

Distal convoluted tubule

Proximal convoluted tubule
Efferent arteriole
Glomerular capsule
Glomerulus
Towards renal vein
From renal artery
Afferent arteriole

Straight venule

Renal medulla

Peritubular capillaries
Straight arteriole
Loop of Henle

Collecting tubule

Peritubular capillaries

Orifice at renal papilla

Arteries, artierioles,
capilleries of glomerulus

Veins, venules

Peritubular capillaries

A

FIGURE 5-2
Diagrams illustrating the arrangement of nephrons and blood vessels in a mammalian kidney.
(A) Two adjacent nephrons and several glomeruli and blood vessels.
(Adapted from Walker, W. F., Jr., and Homberger, D. G. *Vertebrate Dissection*, 8th ed. Philadelphia:
Saunders College Publishing, 1992. Dyce, K. M., Sack, W. O., and Wensing, C. J. G. *Textbook of
Veterinary Anatomy*. Philadelphia: W. B. Saunders Company, 1987.)

*Reproductive
System*

undergo a caudal migration, or **descent**, and come to lie in the scrotum. Testicular blood vessels, nerves, and the sperm ducts are carried back with the testes.

Notice again the testicular arteries and veins (see Exercise 4) extending caudally to pass through a pair of openings in the abdominal wall (Figs. 5-1 and 4-4). Each opening, called the **inguinal canal**, is formed during the descent of a testis. It is not a true perforation of the abdominal wall; rather, it is the site where the muscular and connective tissue layers of the abdominal wall evaginated to contribute to the wall of the scrotum. Carefully remove the skin, if you have not already done so, from the abdominal wall, including the midventral strip of body wall that contains the urinary bladder and penis. Leave the muscle layers of the wall intact. Also remove the skin from the medio-ventral surface of the thighs (Fig. 5-4). As you do this, look for a pair of thin-walled, elongated pouches that extend across the medio-ventral surface of the thigh muscles near the body's midline from the inguinal canals toward the scrotum. These are the cre-

masteric pouches**. Without breaking them open, free the cremasteric pouches from the surrounding structures. The wall of each cremasteric pouch consists of connective tissue, which is an extension of the fasciae of the body wall, and of the **cremasteric muscle**, which is an extension of the obliquus internus and transversus abdominis muscles of the abdominal wall (see Exercise 2). Pass a probe through an inguinal canal and notice that the probe will enter the cremasteric pouch.

Because the cavity of the cremasteric pouch is an extension of the peritoneal cavity, it is lined with serosa and is called the **vaginal cavity**. The vaginal cavity remains connected to the peritoneal cavity in adults of some mammals, such as rats and rabbits, in which the testes can be moved into the scrotum during the breeding season, and returned to the abdominal cavity when the animal is not breeding. However, the testes remain permanently in the scrotum in many adult mammals, such as pigs and human beings, in which the proximal part of the vaginal cavity atrophies.

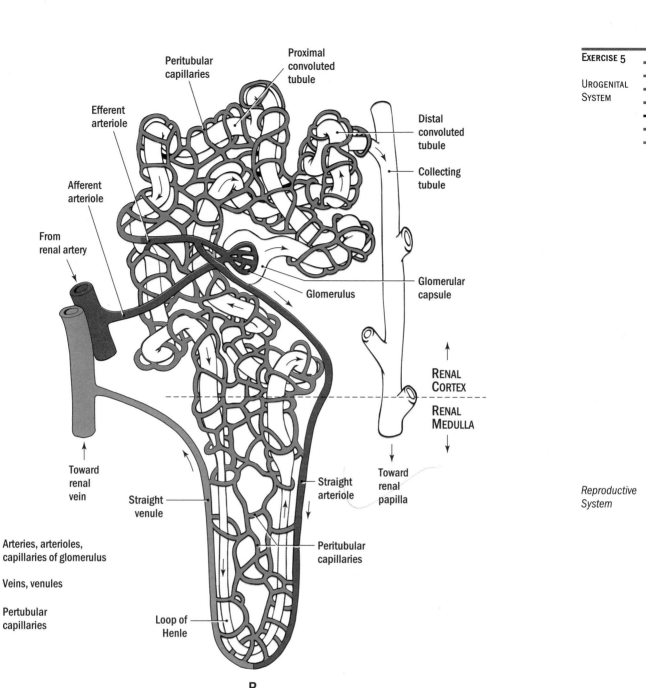

FIGURE 5-2 CONTINUED
Diagrams illustrating the arrangement of nephrons and blood vessels in a mammalian kidney.
(B) Enlargement of a single nephron with its associated capillaries.
Dorit, R. L., Walker, W. F., Jr., and Barnes, R. D. *Zoology*. Philadelphia: Saunders College Publishing, 1991. Stalheim-Smith, A. and
Fitch, G. K. *Understanding Human Anatomy and Physiology*. Minneapolis/St. Paul: West Publishing Company, 1993.

*Reproductive
System*

Leave the inguinal canal and cremasteric pouch intact on one side of the body so that you can demonstrate it to a colleague who is dissecting a female specimen. Open the cremasteric pouch on the other side and find the **testis** (Fig. 5-1). A band of tissue, the **epididymis**, starts at the cranial end of the testis, and extends caudally along one side of the testis to its caudal end, where the epididymis continues into the sperm duct, or **ductus deferens** (deferent duct). Sperm cells are produced in microscopic seminiferous tubules and leave the testis through microscopic ducts to enter the cranial end of the epididymis, where they mature and are stored until ejaculation. The epididymis and ductus deferens have evolved from parts of the ancestral vertebrate kidney system. In fishes and amphibians, sperm cells pass from the testis to certain kidney tubules and, thence, to an archinephric duct that also drains the kidneys. Reptiles, birds, and mammals have evolved a new kidney duct, the ureter. The primitive archinephric duct and some of the ancestral kidney tubules have evolved into the ductus deferens and epididymis of the male genital system.

The serosa of the vaginal cavity is called the tunica vaginalis. The wall of the cremasteric pouch is lined with the **peritoneal tunica vaginalis**, and the testis and associated organs are covered by the **visceral tunica vaginalis**. Notice that the testis, epididymis, ductus deferens, and testicular vessels are anchored by a mesentery, the **mesorchium**, which extends from the dorsal wall of the vaginal cavity to the organs. The cord of tissue that extends from the caudal end of the epididymis to the wall of the vaginal cavity is a thickening of the mesorchium known as the **gubernaculum**. The gubernaculum extends between the testis and the wall of the cremasteric pouch very early in development, and its failure to grow as rapidly as other parts of the body helps to bring about the descent of the testis.

The degree to which the testes have descended will depend on the age of the specimen you are studying. At an early stage of development, the testis may just be passing through the inguinal canal into the cremasteric pouch; at a later stage of development, the testis will already be lodged in the cremasteric pouch. The descent of the testis

into the scrotum is necessary to allow complete maturation of the sperm cells. The final stage of spermatogenesis, in which the spermatids develop tails and lose most of their cytoplasm (Section D of this exercise), cannot be completed if the testes are confined to the abdominal cavity. In most mammals, scrotal temperature is several degrees lower than intraabdominal temperature; the inference is that sperm cell maturation is inhibited by elevated temperatures.

Slightly cranial to the testis, the testicular artery is highly coiled and entwined by a network of testicular veins known as the **pampiniform plexus**. This is another cooling mechanism in which some heat passes from the warmer arterial blood, which is flowing from the core of the body to the testis, to the cooler venous blood returning from the peripheral testis. The ductus deferens, testicular vessels, and an inconspicuous testicular nerve, which are enveloped by the visceral tunica vaginalis, run together between the testis and inguinal canal as a bundle known as the **spermatic cord**. After passing through the inguinal canal, the ductus deferens loops over the ureter and enters the urethra along with the ductus deferens of the opposite side.

The pelvic cavity must be opened to see the rest of the reproductive system. But before this is done, locate the penis in the midventral strip of body wall that contains the urinary bladder. The penis may have been seen between the two cremasteric pouches when they were dissected (Fig. 5-4). Locate again the **preputial orifice** just caudal to the umbilical cord (see Fig. 1-2). The penis extends caudally from here and makes an S-shaped **sigmoid flexure of the penis** near the caudal end of the pelvis. Free the penis

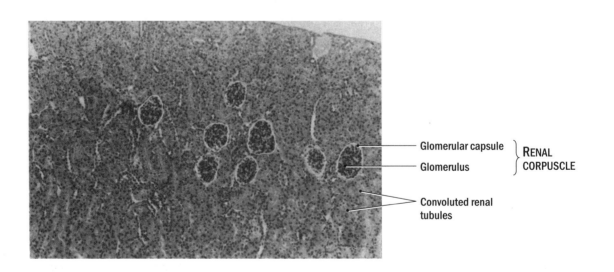

FIGURE 5-3
Photomicrograph through the kidney of a cat.

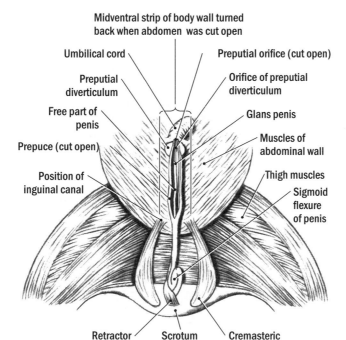

FIGURE 5-4
Ventral view of the penis and cremasteric pouches of a fetal pig. The skin on the abdominal wall and thighs has been removed.

*Reproductive
System*

sufficiently to push it to one side and, with a sharp scalpel, cut through the midventral part of the pelvic muscles and the bony pubic symphysis. Spread the legs apart and expose the pelvic cavity (Fig. 5-1).

Dissecting deep within the pelvic cavity, trace the urethra caudally, separating it from the rectum, or terminal part of the digestive tract. A pair of small glands, the **seminal vesicles**, lies on the dorsal surface of the urethra at the point at which the deferent ducts enter. Turn the urethra to the side sufficiently far to dissect between the seminal vesicles and find the **prostate** (Fig. 5-5). A larger pair of glands, the **bulbourethral glands** (Cowper's glands), flank the urethra near the anus. At the time of ejaculation, all these accessory genital glands secrete a liquid that carries the sperm cells; this liquid is the **seminal fluid**. The seminal fluid also activates and nourishes the sperm cells, and it contains substances that neutralize acidic secretions of the vagina in the female.

Just distal to the bulbourethral glands, the urethra enters the penis, which extends cranially beneath the skin. The base of the penis is enlarged and surrounded by a **bulbocavernous muscle**, which aids in expelling urine and seminal fluid (Fig. 5-5). A slender pair of **retractor muscles of the penis** originates from the sacrum, passes over the caudal surface of the bulbourethral muscle, and

extends cranially along the ventral surface of the penis to insert near the sigmoid flexure of the penis. Cut open the preputial orifice, and the **prepuce**, or penis sheath, which envelopes much of the part of the penis lying in the ventral abdominal wall (Fig. 5-4). The distal end of the penis forms a **glans penis**. Notice the small **orifice of the preputial diverticulum** on the dorsal wall of the prepuce near the preputial orifice. It leads into the **preputial diverticulum**, a pair of small sacs on each side of the cranial end of the prepuce. Some stale urine and decaying sloughed-off cells accumulate in the preputial diverticulum and give adult boars their characteristic odor.

Cut a cross section out of the middle of the penis and observe it under low magnification. The penile part of the urethra is surrounded by a column of spongy, vascular tissue known as the **corpus spongiosum penis**. Two other similar columns of tissue, the **corpora cavernosa penis**, extend along the dorsal surface of the penis. It is not always possible to recognize two distinct columns in this tissue. This spongy, vascular tissue is called **erectile tissue**, because the penis becomes erect when the vascular cavities fill with blood. As the penis of a pig becomes erect, its sigmoid flexure straightens out, and the penis protrudes from the prepuce. The retractor muscles of the penis pull the penis back into the prepuce after an erection and restore the sigmoid

79

*Reproductive
System*

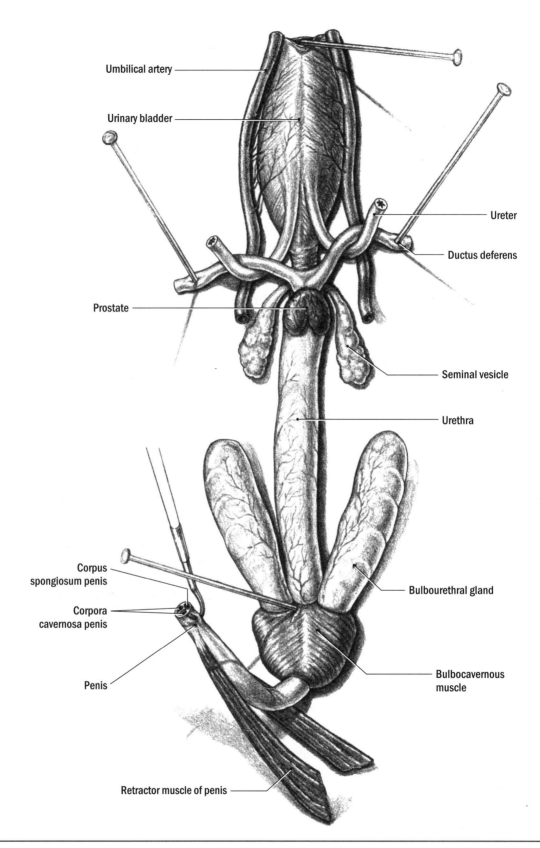

Umbilical artery

Urinary bladder

Ureter

Ductus deferens

Prostate

Seminal vesicle

Urethra

Corpus
spongiosum penis

Bulbourethral gland

Corpora
cavernosa penis

Penis

Bulbocavernous
muscle

Retractor muscle of penis

FIGURE 5-5
Enlarged dorsal view of the male urethra and associated organs of a fetal pig.
The bladder, urethra, and penis are shown as if they had been removed from the body
to show the structures clearly, but you do not need to remove them.

flexure. The penis of a male human being has a similar basic structure, although the prepuce is not as deep, a preputial diverticulum is not present, the distal part of the penis is separated from the ventral abdominal wall, there is no sigmoid flexure, and there are no retractor muscles.

B.2. Female Reproductive Organs

The **ovaries** are a pair of small nodule-like organs located just caudal to the kidneys. Each is anchored to the dorsal body wall by a mesentery, the **broad ligament** (Fig. 5-6). A conspicuous convoluted duct, the **horn of the uterus**, lies in the free edge of the broad ligament. Another cordlike mesentery, the **round ligament**, begins near the ovary, crosses the broad ligament perpendicular to it, and attaches to the abdominal wall near the groin at a point comparable to the location of the male inguinal canal. The round ligament is the female counterpart of the male gubernaculum and helps bring about a partial descent of the ovary during embryonic development.

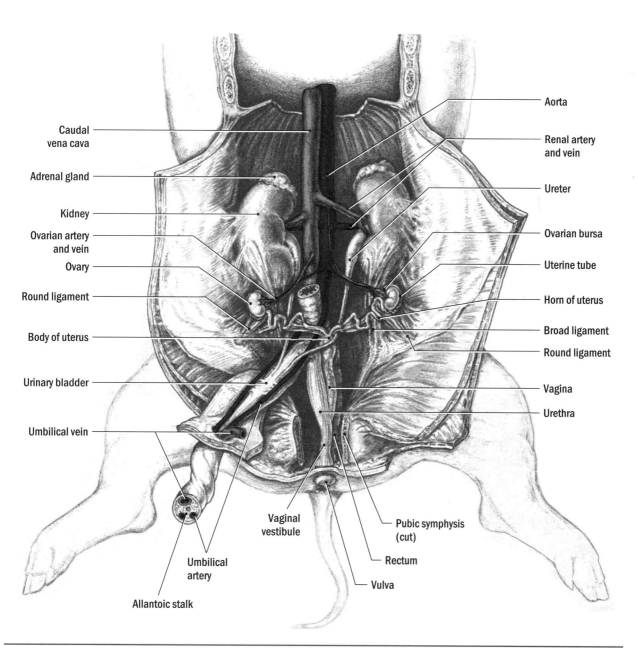

Reproductive System

Caudal vena cava

Adrenal gland

Kidney

Ovarian artery and vein

Ovary

Round ligament

Body of uterus

Urinary bladder

Umbilical vein

Aorta

Renal artery and vein

Ureter

Ovarian bursa

Uterine tube

Horn of uterus

Broad ligament

Round ligament

Vagina

Urethra

Vaginal vestibule

Pubic symphysis (cut)

Rectum

Umbilical artery

Vulva

Allantoic stalk

FIGURE 5-6
Ventral view of the urogenital system of a female fetal pig.

Near the caudal end of the ovary, the uterine horn narrows abruptly to form the small, highly convoluted Fallopian tube, or **uterine tube**, which proceeds to the cranial end of the ovary, lying lateral to the ovary. The uterine tube opens through the **ostium tubae** into the **ovarian bursa**, which is a hoodlike expansion of a part of the broad ligament and covers at least the cranial part of the ovary. You can probe the ovarian bursa with a blunt probe introduced caudally between the uterine tube and ovary. When the mature eggs ovulate, that is, break through the serosa enveloping the ovary, they fall into the ovarian bursa. From here, they are moved by cilia toward the ostium tubae. Fertilization, or the union of sperm and egg, takes place in the uterine tubes.

The uterus is a Y-shaped organ. Trace the uterine horns caudally and observe that they unite to form the **body of the uterus**, which lies dorsal to the urethra in the midline of the body. A uterus of this type, with two horns that enter a median body, is called a **bicornuate uterus**. The pelvic cavity must be cut open to expose the rest of the system. Reflect the skin from the ventral surface of the pelvic region if you have not already done so, and, with a sharp scalpel, cut vertically down along the midventral line through the pelvic muscles and pubic symphysis. Spread the legs apart and open the pelvic cavity. Separate the urethra from the body of the uterus and from the vagina (see later), into which the uterus leads. The vagina unites with the urethra to form a common passage, the urogenital canal, or **vaginal vestibule**, which leads to the body surface. The urogenital opening on the body surface, together with the surrounding skin folds, the **labia**, constitutes the **vulva**.

Separate the uterus, vagina, and vaginal vestibule from the **rectum**, or terminal part of the digestive tract. With a pair of scissors, make cuts that extend from the anus ventrally on each side of the vulva. This procedure will allow you to separate the urogenital organs from the rectum and to turn them over to gain access to the dorsal side of the vagina and uterus. Open the vaginal vestibule, vagina, and uterus by making a longitudinal cut on their dorsal side (Fig. 5-7). A small protuberance, the **glans clitoridis**, will be seen on the ventral surface of the vaginal vestibule near the base of the external **genital papilla**. It is a little mass of erectile tissue corresponding to the male glans penis. Also note the orifice of the urethra into the vaginal vestibule.

The parts of the uterus and vagina can be seen particularly well in this dorsal dissection of the urogenital tract. The horns of the uterus unite to form the body of the uterus, which leads into the neck or **cervix of the uterus**. The cervix of the uterus can be distinguished internally from the vagina by the interdigitating folds that occlude its lumen. The smoother part of the passage that lies between the cervix of the uterus and the vaginal vestibule is the **vagina**.

82

A major difference in the genital system of a female human being is the absence of uterine horns. The uterine tubes lead directly to a median, pear-shaped uterine body. This is a **simplex uterus**. The vagina and urethra do not unite caudally to form a common canal but open independently into a shallow vaginal vestibule between the genital labia.

C. THE FETAL PIG AND ITS EXTRAEMBRYONIC MEMBRANES

Fertilized eggs are carried by currents generated by beating ciliary cells and by peristaltic contractions through the uterine tubes into the horns of the uterus, where embryonic development is completed in 112 to 115 days. If a demonstration of the uterus of a pregnant sow is available, notice that the fetuses tend to be equally spaced in the two horns, and that each fetus causes a local enlargement of the horn. Litter size normally ranges from 6 to 12.

Examine a fetus that has been removed from one of the compartments of a uterine horn. Each fetus is enclosed within an elongated, sausage-shaped **chorionic vesicle** (Fig. 5-8). The surface of the chorionic vesicle bears many folds, including many that can be seen only with magnification. All of them interdigitate with corresponding folds of the uterine lining to form a **diffuse placenta**. In human beings, the placenta forms only over a disk-shaped area of the chorionic vesicle and is called a **discoidal placenta**.

The pig has many round bumps, **areolae**, over the surface of the chorion; usually each one is adjacent to the orifice of a uterine gland.

The placenta is a combination of uterine lining and the wall of the chorionic vesicle. Various degrees of union between the uterine lining and the surface of the chorionic vesicle occur among different groups of mammals. In a fetal pig, the union is not intimate (see Exercise 4, Fig. 4-11). Maternal capillaries in the uterine lining are separated from fetal capillaries in the wall of the chorionic vesicle by four layers of epithelium: the endothelium lining the capillaries of both the maternal and fetal parts of the placenta, and the epithelium lining the uterus and covering the surface of the chorionic vesicle. Nutrients, gases, and waste products diffuse between the fetal and maternal capillaries, crossing the slight space between them. This space is filled with a uterine secretion produced by the uterine glands. In many other mammals, including human beings, there is a more intimate union between the uterine lining and the wall of the chorionic vesicle, because parts of the chorionic vesicle penetrate the uterine lining and maternal blood vessels, and the villi on the surface of the chorion, which contain fetal capillaries, are bathed in pools of maternal blood.

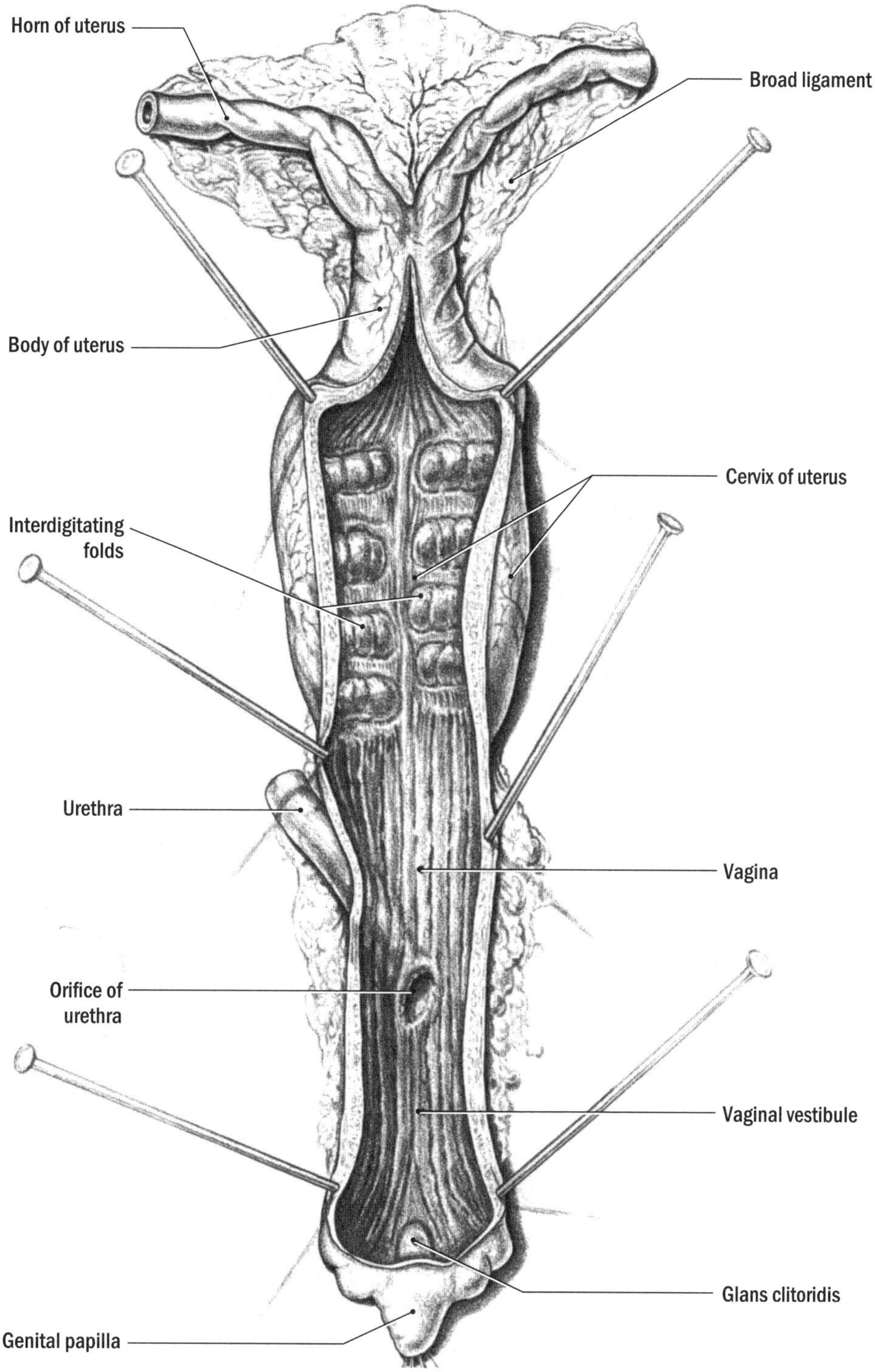

Horn of uterus

Broad ligament

Body of uterus

Cervix of uterus

Interdigitating
folds

Urethra

Vagina

Orifice of
urethra

Vaginal vestibule

Glans clitoridis

Genital papilla

*The Fetal Pig
and Its
Extraembryonic
Membranes*

FIGURE 5-7
Enlarged dorsal view of the uterus, vagina, and vaginal vestibule
of a fetal pig, cut open to show the internal appearance.

Cut open the chorionic vesicle, being careful not to cut through or break a second sac that lies within it and envelopes the fetus. Most of the wall of the chorionic vesicle is composed of a fusion of two **extraembryonic membranes** that develop in association with the fetus. The outer one is the **chorion;** the inner one is the wall, or **allantoic membrane**, of the **allantois**. The allantois is a large sac that grows out from the fetus. Its stalk was seen in the umbilical cord, and the intraembryonic part of it becomes the urinary bladder. When the chorionic vesicle is opened, you are looking into the cavity of the allantois. **Umbilical blood vessels**, which can be seen ramifying in the wall of the chorionic vesicle and entering the **umbilical cord**, lie in the allantoic membrane. The chorion itself is not vascularized. You may notice an undilated part of the chorionic vesicle at each end. The wall of these is composed only of chorion, for the allantois does not penetrate into the ends of the chorionic vesicle.

The fetus is surrounded by another extraembryonic membrane, the thin-walled, nonvascular **amnion. Amniotic fluid** fills the amniotic cavity, prevents adhesions between the fetus and surrounding membranes, acts as a protective water cushion, and is, in a sense, a local aquatic environment in which the buoyant embryo develops. Ancestral vertebrates, of course, developed in water. It was the evolution of the amnion and other extraembryonic membranes that permitted reptiles, birds, and mammals to reproduce in the terrestrial environment. These vertebrate classes are collectively called **amniotes;** amphibians and the several classes of fishes are **anamniotes**.

D. GONADS

The reproductive passages that you have been studying transport gametes and protect and nourish the developing embryo. It is the gonads, the testis and ovary, that produce the gametes. Use histological slide preparations of mammalian gonads to study gamete production.

Gonads

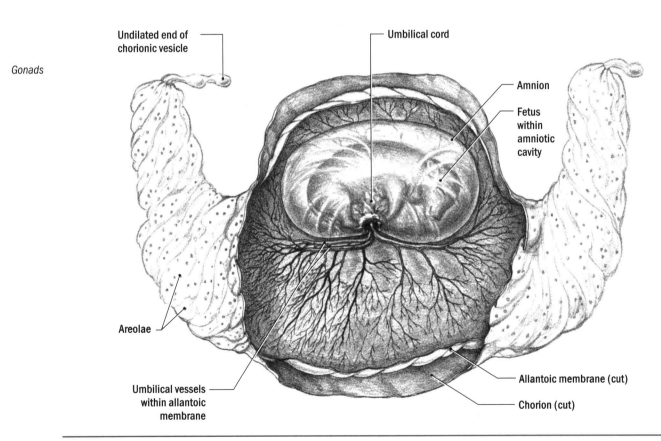

Undilated end of chorionic vesicle

Umbilical cord

Amnion

Fetus within amniotic cavity

Areolae

Umbilical vessels within allantoic membrane

Allantoic membrane (cut)

Chorion (cut)

FIGURE 5-8
Opened chorionic vesicle of a fetal pig. The allantoic membrane is closely apposed to the chorion.
Part of the allantoic membrane is peeled away from the chorion in this dissection.

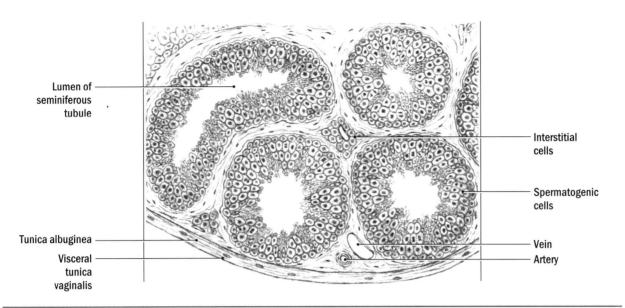

Lumen of
seminiferous
tubule

Interstitial
cells

Spermatogenic
cells

Tunica albuginea

Visceral
tunica
vaginalis

Vein

Artery

FIGURE 5-9
Diagrammatic low magnification of part of a mammalian testis, showing several seminiferous tubules.

D.1. Testis

The **testis** is composed primarily of many long and highly coiled **seminiferous tubules,** of which some will be seen cut in cross section, others in oblique section, and others tangentially (Fig. 5-9). The total length of all of them in one testis of a male human being is estimated to be about 250 meters. A tough, fibrous capsule, the **tunica albuginea,** forms the wall of the testis and sends septa into the organ dividing it into many lobules. A thin serosa, the visceral tunica vaginalis, envelopes the testis.

Examine the walls of several seminiferous tubules (Fig. 5-10). Chromosomes are visible in many of the cells, for they are dividing rapidly. The periphery of the tubule wall contains **spermatogonia,** which start to divide mitotically at puberty. Some of their daughter cells remain spermatogonia and stay near the periphery of the tubule; others enlarge, move toward the lumen, and become **primary spermatocytes.** Spermatogonia and primary spermatocytes contain the diploid number of chromosomes found in body cells. Each primary spermatocyte undergoes the first meiotic division to form two **secondary spermatocytes,** and each of these divides in the second meiotic division to form two **spermatids.** Thus, four haploid spermatids develop from a single diploid primary spermatocyte. Cells do not grow between successive cell divisions, so they become progressively smaller as they move closer to the lumen of the seminiferous tubule. A spermatid undergoes a transformation into a mature,

motile **spermatozoon.** Its nucleus condenses and appears very small and dark, much of the cytoplasm is lost, and a tail develops. The various stages of spermatogenesis are most easily identified by relative cell size and position within the wall of the tubule.

The spermatogenic cells are closely associated with large supporting **Sertoli cells.** The Sertoli cells form the supporting framework of the tubule wall. Their outlines will not be clear because their cell membranes can only be seen in electron micrographs (Fig. 5-10A and B). The expanded bases of the Sertoli cells are firmly united with each other at the periphery of the tubule, and their cell bodies extend into the lumen of the tubule between groups of spermatogenic cells. Clusters of spermatogenic cells mature in recesses between and within Sertoli cells, which probably have a nutritive as well as a supporting role.

Successive waves of spermatogenesis travel along the seminiferous tubules; consequently, at any given cross-sectional level of the testis, groups of spermatogenic cells in different tubules will probably be in different stages of spermatogenesis. Some groups may be in early stages of spermatogenesis; others, in the spermatid and spermatozoa stages. A cycle of spermatogenesis from spermatogonia to spermatozoa lasts 16 days in human beings. Vast numbers of spermatozoa are produced, as many as 200 to 300 million per ejaculation.

Small clusters of relatively large, light-staining **interstitial cells** are present in groups in the connective tissue between the seminiferous tubules (Figs. 5-9 and 5-10B).

Gonads

85

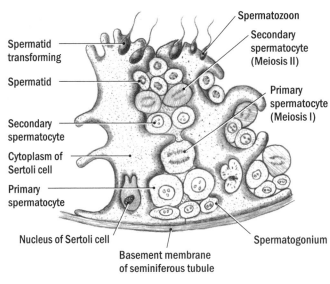

Lumen of seminiferous tubule

Spermatid transforming

Spermatid

Secondary spermatocyte

Cytoplasm of Sertoli cell

Primary spermatocyte

Nucleus of Sertoli cell

Spermatozoon

Secondary spermatocyte (Meiosis II)

Primary spermatocyte (Meiosis I)

Spermatogonium

Basement membrane of seminiferous tubule

A

Gonads

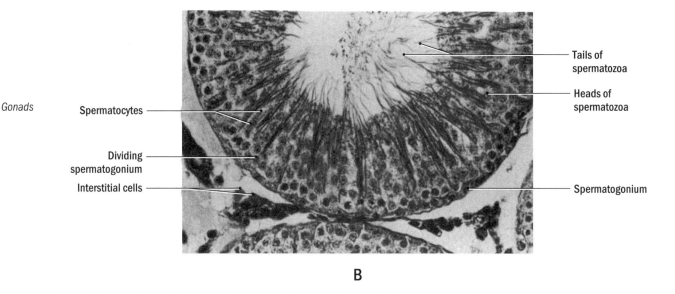

Spermatocytes

Dividing spermatogonium

Interstitial cells

Tails of spermatozoa

Heads of spermatozoa

Spermatogonium

B

FIGURE 5-10
Structure of a seminiferous tubule. (A) Diagram of part of the wall of a seminiferous tubule,
illustrating the close relationship between maturing spermatogenic cells and Sertoli cells.
The cell boundaries of Sertoli cells can only be seen in electron micrographs.
This is a composite diagram showing all of the stages in spermatogenesis.
All stages would not be seen at a single level of the tubule.
(B) Photomicrograph of part of the seminiferous tubule of a rat.

They produce **testosterone**, the male hormone responsible for the enlargement of reproductive passages and glands, for the development of male secondary genital characteristics, and for male sexual behavior. The development of the seminiferous tubules and the production of testosterone are controlled by the same pituitary **gonadotropic hor-**

mones that regulate the ovarian cycle in females. **Follicle stimulating hormone** (FSH) is necessary for the development of the seminiferous tubules and Sertoli cells, and **luteinizing hormone** (LH) is needed by the interstitial cells to synthesize testosterone. LH also acts with testosterone in promoting spermatogenesis. Spermatogenesis is seasonal in seasonally breeding mammals but is continuous during the adult life of male human beings.

D.2. Ovary

In contrast to the testis, which consists essentially of a mass of tubules, the ovary consists of a solid mass of cells and tissues. Examine a slide and notice the small **ovarian medulla** in the center of the ovary, which is composed of a highly vascular connective tissue. The thicker **ovarian cortex** consists of a denser connective tissue containing developing follicles with maturing eggs (Figs. 5-11 and 5-12). As was the case in the testis, the cortical connective tissue forms a thin, very dense layer around the organ, which is devoid of follicles and is called the **tunica albuginea**. The surface of the ovary is covered by a thin serosa, or visceral peritoneum. In the testis, the maturation of spermatogonia into mature sperm is a continuous process that can be reconstructed by observing sections of several seminiferous tubules. However, the maturation of an egg and follicle cannot be observed in a slide of one ovary, because the maturation of the egg and follicle takes place in a stepwise fashion at different stages in the life of a female.

During embryonic development of a female, the diploid **oogonia** multiply by mitosis, start their first meiotic divisions (to the prophase stage), and mature into diploid primary oocytes. The primary oocytes are surrounded by a layer of thin follicular, or granulosa, cells. The primary oocyte and surrounding cells together form a primordial follicle. The oogonia stop dividing at birth, by which time they number between 200,000 and 500,000 per ovary in a female human being. This number is more than enough to last through the reproductive years of an individual, during which usually only one mature egg is released during each menstrual cycle. Tens of thousands of primary oocytes atrophy without ever becoming fully mature.

The primordial follicles remain dormant until puberty. Beginning at sexual maturity, the follicles and maturing eggs undergo recurrent **ovarian cycles**, during which the follicles enlarge and the eggs mature and are discharged from the ovary. The number of eggs released and the cycle duration and frequency varies with the species. Deer and other seasonal breeders have one or more ovarian cycles during a brief reproductive season, and the ovary is dormant most of the year. Cats, dogs, and

Gonads

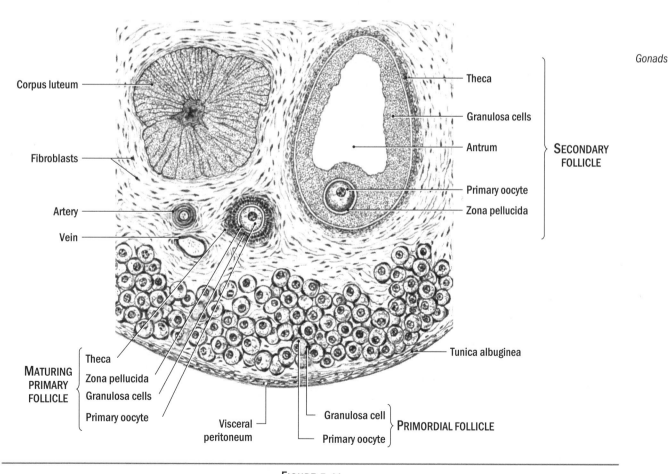

FIGURE 5-11
Diagrammatic microscopic section of part of the ovarian cortex from a cat.

many of our domestic mammals, have two or more reproductive periods per year. Many other mammals, including rats and human beings, have repetitive ovarian cycles throughout the year. The human ovarian cycle is 28 days long, and it is accompanied by a **menstrual cycle**, during which the uterine lining undergoes changes for the reception of a fertilized egg and sloughs off if one does not reach the uterus.

Look for **primordial follicles** in the periphery of the ovarian cortex just below the tunica albuginea (Figs. 5-11 and 5-12). They may be very numerous. Each primordial follicle contains a **primary oocyte** and a single layer of flattened epithelial cells, the **granulosa cells**. A primary oocyte is characterized by a conspicuous, eccentric nucleus embedded in a relatively clear cytoplasm.

At the beginning of an ovarian cycle, several primordial follicles start to mature into **primary follicles**, which may be seen deeper within the ovarian cortex. The granulosa cells multiply by mitosis, assume a cuboidal shape, and form two or more layers around the primary oocyte. Cells derived from the surrounding connective tissue envelop the granulosa cells and form a envelope, or

Gonads

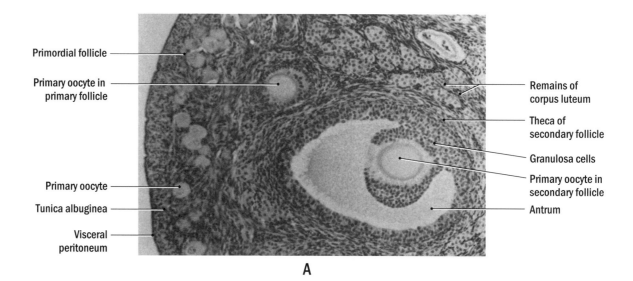

A

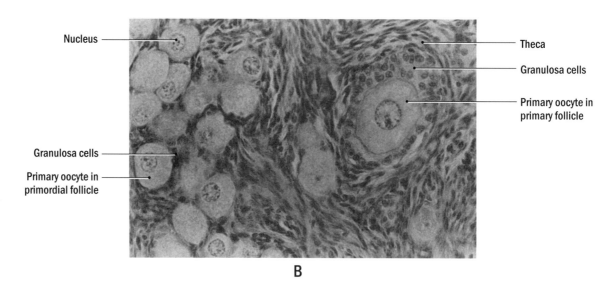

B

FIGURE 5-12
Photomicrographs of the ovarian cortex of a cat.
(A) At low magnification.
(B) At high magnification.

theca, for the primary follicle. The primary oocyte enlarges as food reserves accumulate in its cytoplasm, and it becomes surrounded by a lighter-staining region known as the **zona pellucida**.

As the ovarian cycle continues, certain primary follicles enlarge and become oval in shape. In each enlarging follicle, a space, or **antrum**, develops amid the granulosa cells and becomes filled with a follicular liquid. The primary oocyte is located near one pole of the follicle, which is now called a **secondary follicle**.

Before ovulation, growth accelerates in one (in a female human being) of the secondary follicles. It becomes very large and moves toward the surface of the ovary. It is usually only this one follicle that becomes a mature **tertiary**, or Graafian, **follicle**. The primary oocyte now completes the first meiotic division, generating a **secondary oocyte** containing nearly all of the cytoplasm and a small **first polar body**, which has little cytoplasm and is destined to disintegrate. It is unlikely that you will see polar bodies.

A rapid accumulation of follicular liquid and increased pressure within the antrum leads to **ovulation**. The follicle and ovary surface rupture; and the secondary oocyte, with a few surrounding granulosa cells, is discharged into the part of the coelom adjacent to the entrance of the uterine tube. As the secondary oocyte enters the uterine tube, it begins its second meiotic division, but this division is not completed and remains in the metaphase stage until a sperm penetrates the egg. At sperm penetration, which must occur in the uterine tube within 24 to 72 hours in a female human being, the secondary oocyte completes the second meiotic division, forming a mature haploid **ootid** (ovum) containing nearly all of the cytoplasm and the **second polar body**, which will soon die and be reabsorbed. Fertilization is completed by the union of male and female nuclear materials. Again the development of male and female gametes differs. In the testis, four functional speramatozoa develop from each primary spermatocyte; in the ovary, only a single ootid develops from a primary oocyte. This developmental pattern is an adapta-

tion that conserves all of the cytoplasm and nutrients in one cell and provides the fertilized egg with sufficient reserves to travel through the uterine tube and implant in the uterine lining (about 7 days in a female human being). A placenta develops soon after implantation.

After ovulation, most of the cells of the tertiary follicle remaining in the ovary are transformed into a **corpus luteum**, which may be seen in some slides (Fig. 5-11). The granulosa and inner thecal cells enlarge and accumulate lipids and other materials. The human corpus luteum lasts until shortly before the onset of the next menstruation and then regresses. This regression marks the end of an ovarian cycle. If a pregnancy ensues, the corpus luteum enlarges more and lasts well into the pregnancy.

The follicles not only protect and nourish the developing eggs, they and the corpora lutea also are endocrine glands. The follicles develop under the influence of **follicle stimulating hormone** secreted by the anterior part of the hypophysis, or pituitary gland (Exercise 7). As the follicles mature, they secrete **estrogen**, which is responsible for the enlargement of the uterus, for the development of the female secondary genital characteristics, and for the buildup of the uterine lining following a menstrual period. The corpora lutea develop under the influence of another hypophyseal gonadotropin, **luteinizing hormone**. Some estrogen is produced by the corpora lutea, but their primary hormone is **progesterone**. Progesterone is needed for the final development and increased vascularization of the uterine lining in preparation for the implantation of an embryo and for the maintenance of the uterine lining and placenta during the early stages of pregnancy. Blood levels of estrogen and progesterone affect the release of gonadotropic releasing and inhibiting hormones from the hypothalamus of the brain, and these hormones control the release of the gonadotropic hormones. A complex and interwoven feedback mechanism among the hormones of the hypothalamus, hypophysis, and ovary controls the ovarian cycle, time of ovulation, changes in the uterine lining, pregnancy, and birth.

Gonads

EXERCISE

SIX

Nervous Coordination: Sense Organs

S URVIVAL OF ANY ORGANISM requires the harmonious interaction of its various organs and appropriate responses of the entire organism to changes in the external and internal environments conditions. Many aspects of metabolism, growth, and reproduction are regulated by the secretions of the numerous endocrine glands. (The thyroid gland, islets of Langerhans, and adrenal gland have been mentioned in Exercises 3 and 4; the hypophysis and pineal body will be considered during the study of the brain in Exercise 7.) The circulatory system carries the hormones secreted by an endocrine gland in one part of the body to other parts of the body. Some hormones have a broad effect and affect most cells in the body, but other hormones have a more narrow effect on certain cells that bear specific receptor molecules for particular hormones. Endocrine integration tends to be slow, long lasting, and widespread in its effects. Feedback mechanisms of various types control the activity of the endocrine glands and the hormonal levels in the blood.

More rapid and specific control and integration of the internal organs, as well as of most of an animal's responses to the external environment, is mediated through the receptor organs, the nervous system, and effector organs such as muscles, glands, and ciliated epithelia. **Neurons**, or nerve cells, which are the basic cellular units of the nervous system, may be activated by nearly any stimulus that is intense enough. The specificity of nervous integration derives from the facts that the receptor cells of the body are attuned to specific environmental parameters (light, temperature, mechanical forces, chemical substances) and that they are activated by very slight changes in these parameters. The receptor cells, in turn, activate the neurons with which they are connected. These neurons have specific interconnections with other neurons and, eventually, with appropriate effectors.

Although distinct in many ways, hormonal and nervous integration mechanisms have much in common and tend to interact with each other. Both use chemical messengers either to carry information through the circulatory system or across the synapses between neurons. Parts of the brain, as we shall see, influence the endocrine system, and the level of hormones in the blood affects many behavioral responses controlled by the nervous system.

We will examine receptors in this exercise and the nervous system in Exercise 7. The actual receptor cells may be modified neurons, such as the olfactory cells of the nose, or they may be specialized cells, such as the rods and cones in the retina of the eye. It is convenient to classify receptor cells according to the environmental stimulus to which they are particularly sensitive. **Chemoreceptors** in the nose and on the tongue respond to slight changes in the chemical environment; **mechanoreceptors** in the skin, muscles, and ear respond to slight mechanical deformations in tissues; **photoreceptors** in the retina of the eye respond to changes in the intensity and wavelength of light; and **thermoreceptors** respond to changes in temperature and infrared radiation. Receptor cells by themselves cannot be seen without high magnification, but many of these cells aggregate along with other tissues to form grossly visible **sense organs**, such as the eye, ear, and nose. The associated tissues support and protect the receptor cells and may amplify the environmental stimulus and direct it toward the receptors.

A. EYE

The eye is a complex sense organ attuned to light and to changes in the visual field. The sense organ itself, the **eyeball**, is lodged within a socket of the skull, the **orbit**, and is surrounded by various accessory structures. Notice again the **upper** and **lower eyelids**. They are movable in terrestrial vertebrates, and they protect the eyeball and help to keep its surface moist by spreading tears across it. Make an incision extending forward from the rostral corner of the eye, if you have not already done so in Exercise 1, and notice the **nictitating membrane**, or third eyelid. The nictitating membrane can move across the surface of the eyeball in most mammals and, thus, helps to keep it clean. In human beings, it is reduced to a small tissue fold (the semilunar fold) in the medial corner of the eye.

Carefully dissect away the eyelids, noticing, as you do so, that the delicate membrane lining the inner surface of the eyelids, the **conjunctiva**, reflects over the surface of the eyeball where it fuses with the cornea (see later). With a strong pair of scissors, remove some of the bone surrounding the eyeball so that you can see the eyeball more clearly. Tissue around the eyeball should be carefully picked away. Much of this tissue is simply connective tissue, but some of it has a glandular texture. These are the tear glands, or **lacrimal glands**, which continuously produce a watery secretion that flows over the surface of the eyeball, moistening and cleaning it. Tears are drained by a **nasolacrimal duct** into the nasal passages. (The duct is difficult to find in the fetal pig, but one entrance into it, the **lacrimal punctum**, can be seen in your neighbor's eye by pulling down the lower eyelid and looking for a small pore on the edge of the eyelid near the most medial eyelash. A similar pore is located on the upper eyelid.)

Narrow, bandlike muscles will be seen extending from deep within the orbit to attach onto the surface of the eyeball. These are the **extrinsic ocular muscles**, which control the movements of the eyeball as a whole. There are seven of them, but you may not find all of them. The two eyeballs must move in unison, for example, when you follow these words across the page or follow a moving object. As you pick these muscles away, you will find the **optic nerve** emerging from the skull and entering the deep surface of the eyeball. Bisect the optic nerve and remove the eyeball from the orbit.

Dissect the eyeball of your specimen of a fetal pig or a larger eyeball from a sheep or cow. If you study a larger eyeball, first remove the fat and extrinsic ocular muscles and find the optic nerve, which attaches to the medial side of the eyeball slightly rostral and ventral to its center. This will help you orient the specimen in space. The eyeball is composed of three layers: fibrous tunic, vascular tunic, and retina. The outermost, a dense fibrous layer, is the **fibrous tunic**. The posterolateral (deep) part of the fibrous tunic is an opaque **sclera**; but at the front of the eyeball, the fibrous tunic is a transparent **cornea**. Although the cornea is rather cloudy in preserved specimens, you can usually look through it and see the pigmented iris, with a circular opening, called the pupil, in its center (see also later).

With a pair of fine scissors, cut off approximately the dorsal one-quarter of the eyeball, including the top of the cornea and iris. Continue cutting through the sclera. Before dissecting further, look into the eyeball. Note again the outer fibrous tunic (Fig. 6-1). A dark **choroid**, a part of the **vascular tunic**, lies internal to the fibrous tunic, and the whitish **retina** lies internal to the choroid. Light entering through the pupil falls on this layer. The retina is not attached to the vascular tunic but is held against it by the gelatinous **vitreous body**, which fills the large space between the **lens** near the front of the eyeball and the back wall of the eyeball. If you are dissecting a fetal eye, notice the **hyaloid artery**, which enters the eyeball with the optic nerve and passes through the vitreous body to nourish the developing lens. It disappears at about the time of birth. Its continued presence would cast a shadow on the retina.

Before dissecting further, submerge the eyeball in a dish of water so that delicate structures will be supported. Now make a vertical, equatorial cut around the center of the eyeball in the plane of the dashed arrows shown in Fig. 6-1. This incision will divide the eyeball into front, or anterior, and back, or posterior, halves (Fig. 6-2). Carefully remove the vitreous body, and you will see the retina in the posterior half of the eyeball more clearly (Fig. 6-2A). It may have floated away from the choroid. The whitish or grayish layer you are looking at is the **nervous layer of the retina**. Embryonically, there is also a **pigmented layer of the retina**, but this becomes attached to the choroid in the adult. Because the retina develops as an outgrowth of the embryonic brain, its nervous layer contains several layers of neurons, with complex interconnections between them.

The photoreceptive cells, the **rods** and **cones**, lie in the retina next to the choroid, consequently light entering the eye must pass first through the neuronal layers to reach and stimulate them. Rods are activated by weak light and allow the animal to see in dimly lighted areas; cones are activated by brighter light. Because there are three kinds of cones, with light absorption maxima in the red, green, and blue parts of the light spectrum, cones also mediate color vision. Considerable processing of the image occurs within the retina before neuronal signals continue along the optic nerve to the brain.

The retina is firmly attached to the wall of the eyeball only at a small circular area, the **optic disk**, where its

Eye

neurons leave the retina to form the optic nerve. Because rods and cones are absent in this area, the optic disk is a "blind spot," and an image that falls on this area cannot be perceived.

Let the retina float away from the wall of the eyeball. Most of the vascular tunic is the black choroid, which you have been looking at. It lies between the sclera and the retina. The choroid is rich in blood vessels that help nourish the retina. Its pigment, which is derived from the pigmented layer of the embryonic retina, absorbs scattered light (like the black color inside a camera) and, thus, prevents a blurring of the image falling on the retina. In the eyes of some mammals, such as sheep and cows, but not pigs and human beings, part of the choroid is modified and has an iridescent sheen. This area is the **tapetum lucidum**. It reflects some of the light passing through the retina back onto the photoreceptive cells, thereby facilitating vision under dim light condi-tions. This reflected light causes the "eye shine" that you may have seen when car lights shine on an animal in the dark.

Now study the anterior half of the eyeball (Figs. 6-1 and 6-2B). The choroid and a part of the retina extend toward the **lens**. The lens is held within a ring of the vascular tunic known as the **ciliary body**. The ciliary body has a somewhat pleated appearance. Carefully remove the lens, noticing that it tends to adhere to the ciliary body because microscopic **zonule fibers** pass from the ciliary body to the periphery of the lens. The oval lens of a preserved specimen has lost the clarity and elasticity that it had in life, but try passing it over the words on this page. You may see that they are magnified slightly because the lens bends, or **refracts**, the light waves.

Now you can see the **iris**, which is an extension of the vascular tunic in front of the lens, and the opening, the **pupil**, in the center of the iris. The space between the iris

Eye

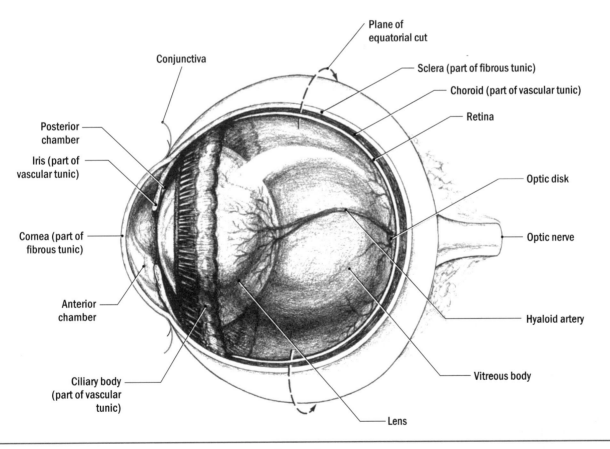

FIGURE 6-1
Eyeball of a fetal pig. You are looking into the eyeball from above after a tangential slice has been removed from the top. The second cut should be made in the equatorial plane, as shown by the dashed arrows.

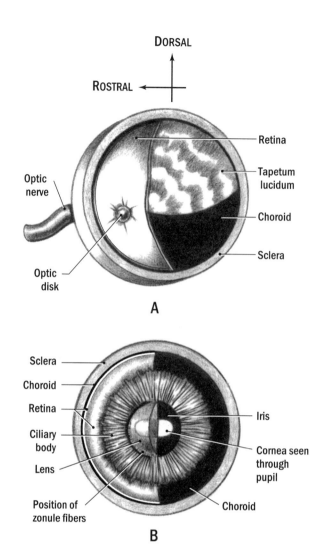

DORSAL

ROSTRAL ←

Retina

Optic
nerve

Tapetum
lucidum

Choroid

Sclera

Optic
disk

A

Sclera

Choroid

Retina

Ciliary
body

Lens

Iris

Cornea seen
through
pupil

Position of
zonule fibers

Choroid

B

Eye

FIGURE 6-2
Drawings of the inside of a sheep eyeball.
(A) Posterior half after removal of the vitreous body as seen from the front of the eyeball.
(B) Anterior half after removal of the lens as seen from the back of the eyeball.
A portion of the retina has been removed in each half to show the underlying vascular tunic.
(Redrawn from W. F. Walker, Jr. and D. G. Homberger, *Vertebrate Dissection*,
8th ed. Philadelphia: Saunders College Publishing, 1992.)

and the place where the lens you removed was located is the **posterior chamber**; the space between the iris and the cornea is the **anterior chamber**. Both of these chambers are filled with a watery **aqueous humor** during life. This liquid is secreted by the ciliary body and maintains a certain intraocular pressure. Surplus aqueous humor is drained through a microscopic canal at the periphery of the cornea.

Light enters the eyeball through the cornea. Because the cornea has a refractive index that is considerably greater than that of the surrounding air, the light waves passing through it are slowed down and refracted toward the optic axis of the eyeball. The amount of light that continues through the eyeball is determined by the diameter of the pupil, and this is controlled by radial and circular smooth muscles in the iris. Light continues through the lens, which refracts it more, thereby causing a sharp, but inverted, image to fall on the retina.

The lens resembles the fine adjustment on a microscope or camera, because it brings the image into sharp focus. However, unlike the process of focusing a microscope or camera, focusing in mammals occurs not by moving the lens back and forth, but by changing its shape. When the eye is at rest, the intraocular pressure produced by the aqueous humor pushes the wall of the eyeball outward near the lens. This places the lens under tension, because it is attached by the zonule fibers to the ciliary body and wall of the eyeball. The elastic lens is somewhat flattened, and distant objects are in focus. To focus, or **accommodate**, for near objects, circular and longitudinal smooth muscle fibers within the ciliary body contract. This action shortens the distance between the ciliary body and the lens, tension on the lens is reduced, and the elastic recoil of the lens causes it to bulge slightly. Its increased thickness increases the refraction of light, and close objects are brought into focus.

The human eyeball is structurally and functionally very similar to the one you have been studying. The lens loses its elasticity with age, and older people often are not able to focus on near objects without glasses. The lens may also become cloudy with age, giving rise to a cataract; fortunately, the lens can be replaced.

B. EAR

The mammalian ear has a dual function—sensing changes in the body's position (equilibrium) and detecting sound. It consists of three parts: **inner ear, middle ear,** and **external ear.** Its receptive cells lie within liquid-filled sacs and ducts of the inner ear, which are embedded deep within the otic capsules of the skull (Exercise 1). It is impractical to dissect the inner ear, which is best examined on models or demonstrations, if they are available. The semicircular ducts, utriculus, and sacculus of the inner ear (collectively called the vesibular apparatus) are related to the sense of equilibrium. Hairlike cytoplasmic processes of the receptive cells are displaced by movements of the head, so an animal can detect changes in body position and movement. Similar receptive cells in the coiled, snaillike cochlea are activated by pressure waves that are generated in the liquid within the cochlea by sound waves. External and middle ears gather and amplify the airborne sound waves sufficiently to generate pressure waves in the cochlear liquid.

The most conspicuous part of the external ear is the large flap, or **auricle**, which, like an old-fashioned ear trumpet, gathers and focuses the sound waves on the eardrum. The human auricle does not normally move, but you may have noticed on a dog or cat that the auricle is turned and directed toward the source of a sound the

animal is interested in. A canal, called the **external acoustic meatus**, extends ventrally and medially from the base of the auricle to the eardrum, or tympanic membrane.

Cut off the auricle and cut away muscle tissue to expose the skull ventral to the external acoustic meatus (Fig. 6-3). Follow the external acoustic meatus toward the skull, carefully cutting it away as you do so. It turns ventrally and becomes encased by bone, which can be picked away with strong forceps. Remove enough tissue to expose the rather large **tympanic membrane.** You may have to cut away some of the lower jaw to see the tympanic membrane clearly.

The tympanic membrane separates the external ear from the middle ear, which consists of a relatively large middle ear cavity, or **tympanic cavity**, and three small bones, the **auditory ossicles**, or ear ossicles, which transmit movements of the tympanic membrane to the inner ear. The rest of the dissection should be done with the aid of a magnifying lamp or dissecting microscope. A little bar of bone, which is part of the malleus, the most lateral of the auditory ossicles, can be seen through the translucent tympanic membrane (Fig. 6-4).

Pick away the tympanic membrane, being careful not to dislodge the malleus, and pick away a bit of the skull just dorsal to the malleus to reveal the tympanic cavity. The slit in the rostroventral portion of the tympanic cavity is part of the **auditory tube** (Eustachian tube). Pass a probe through the auditory tube and notice that it opens into the nasal pharynx (see Exercise 3). The auditory tube permits the equalization of pressure on each side of the tympanic membrane, thereby preventing the tympanic membrane from being subjected to tension due to unequal air pressure on both sides of the tympanic membrane and allowing it to vibrate freely.

The three auditory ossicles (the **malleus, incus,** and **stapes**) are situated in the dorsal portion of the tympanic cavity. The **chorda tympani**, a branch of the facial nerve supplying certain salivary glands and taste buds on the tongue, crosses the malleus. The innermost ossicle, the stapes, fits into a small hole called the oval window, or **fenestra vestibuli,** which is located in the plate of bone forming the medial wall of the tympanic cavity. The fenestra vestibuli leads to the cochlea of the inner ear.

The chain of auditory ossicles transforms the movements of the tympanic membrane (in response to sound waves) into pressure waves in the liquid of the cochlea. The sound waves that are generated in air attenuate quickly as they travel through air and must be amplified to induce pressure waves in the denser liquid of the cochlea. Amplification is accomplished by concentrating most of the energy that impinges on the rather large tympanic membrane on the relatively small membrane in the fenestra vestibuli through the auditory ossicles. The three auditory ossicles together form an articulated chain of

Ear

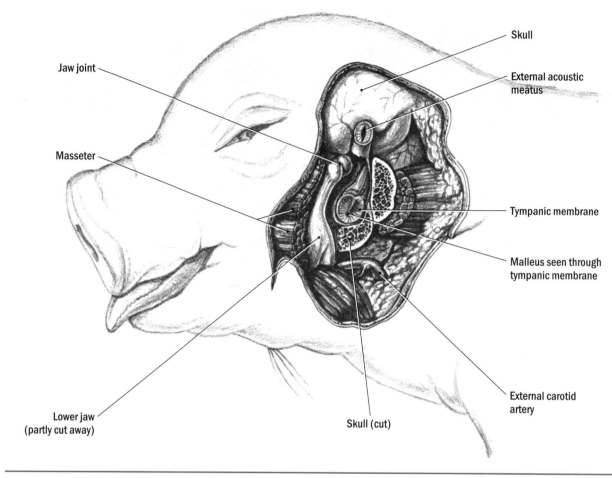

Jaw joint

Masseter

Lower jaw
(partly cut away)

Nose

Skull

External acoustic
meatus

Tympanic membrane

Malleus seen through
tympanic membrane

External carotid
artery

Skull (cut)

FIGURE 6-3
Dissection of the head of a fetal pig to show the location of the tympanic membrane.
Bone surrounding much of the external acoustic meatus has been cut away.

ossicles, which can be stiffened by the action of two small and inconspicuous muscles in the tympanic cavity. Stiffening the chain of ossicles is a reaction to loud noises. As a consequence, the vibrations of the tympanic membrane are reduced and the energy reaching the fenestra vestibuli is lessened. A very loud, intense sound tends to cause damage to the hearing cells in the inner ear.

Slightly ventral and posterior to the fenestra vestibuli, you will see a round opening—the round window, or **fenestra cochleae**—through which pressure waves are released from the cochlea back into the tympanic cavity. Carefully remove the ear ossicles, let them dry, and try to identify them. The small incus lies between the malleus and stapes. A glimpse of the **cochlea** may be had by carefully removing some of the bone forming the medial wall of the tympanic cavity.

The human middle ear is essentially the same as the **96** one of the fetal pig.

C. NOSE

Parts of the nose were studied with the respiratory system on demonstration sagittal sections of the head (Fig. 3-3). Dissect the nose on your specimen at this time. Open the left **nasal cavity** by making a longitudinal section through the snout parallel to the sagittal plane. The section should pass through the left **external nostril** (*naris*). Do not extend the cut into the cranial cavity if you are going to study the brain on your own specimen. Pick away the folds of tissue in the left nasal cavity until you reach the cartilaginous **nasal septum** that separates left and right nasal cavities.

As you expose the nasal septum, look for the **vomeronasal organ** near the base of the rostral part of the septum (Fig. 6-5). It is a small, tubular sac that connects with the roof of the mouth through a small **incisive duct** (see

Exercise 3). The organ is supplied by a special branch of the olfactory nerve. The vomeronasal organ of many mammals is sensitive to **pheromones**, or chemical markers deposited in the environment by other members of the same species. In particular, it is sensitive to pheromones important in social and sexual interactions between members of the same species. Pheromones are used to mark territories, to indicate social status and sexual state, and to convey other social messages. Although this organ starts to develop in the human fetus, it is not certain that it persists as a functional organ in adults. It is smaller in pigs than in most other mammals, and it is absent in aquatic mammals, such whales and porpoises.

Carefully cut away the nasal septum and expose the contents of the right nasal cavity. The cavity is largely filled with folds of tissue, the **nasal conchae**, or turbinates, whose pattern is shown in Figure 6-5. The air passages between them are the **nasal meatuses**. The meatuses converge caudally and enter the nasal pharynx by way of the **internal nostril** (*choana*). The nasal conchae increase the surface area available for olfaction and conditioning the respiratory air.

If your specimen is well injected, you can see the extensive vascular network in the mucous membrane of the nasal conchae. As inspired air crosses the conchae, it is warmed and moistened, and dust particles are

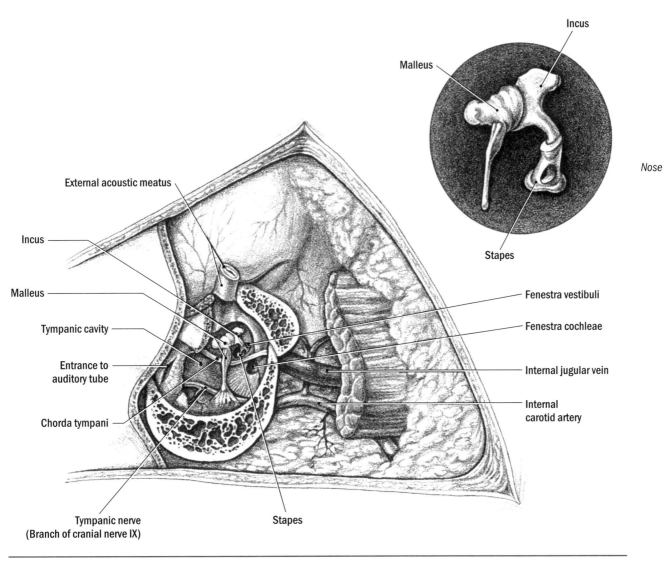

FIGURE 6-4
Dissection of the tympanic cavity of the fetal pig. *Inset*: Enlargement of the auditory ossicles.

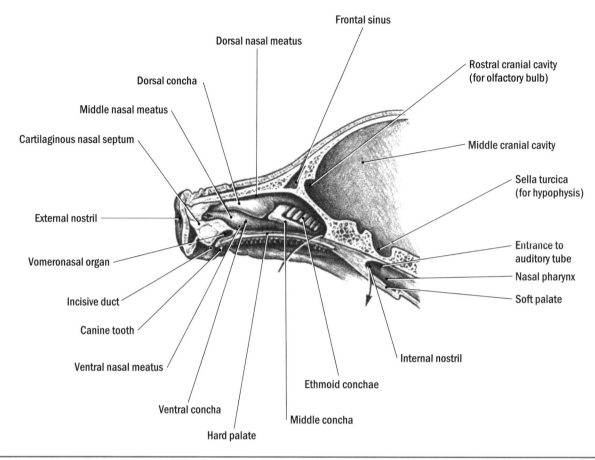

Dorsal nasal meatus

Frontal sinus

Dorsal concha

Rostral cranial cavity
(for olfactory bulb)

Middle nasal meatus

Cartilaginous nasal septum

Middle cranial cavity

External nostril

Sella turcica
(for hypophysis)

Vomeronasal organ

Entrance to
auditory tube

Incisive duct

Nasal pharynx

Canine tooth

Soft palate

Ventral nasal meatus

Internal nostril

Ventral concha

Ethmoid conchae

Hard palate

Middle concha

Nose

FIGURE 6-5
Left vomeronasal organ and right nasal cavity of a fetal pig, as seen in a longitudinal section.

entrapped in mucus secreted by the mucous membrane. As moist air from the lungs is expired, some of the water within it condenses on the mucous membrane and is reabsorbed.

The ethmoid conchae lie just in front of the rostral cranial cavity, which contains the olfactory bulb of the brain. Branches of the olfactory nerve ramify over the con-chae and, in particular, the ethmoid conchae. If you dissect them carefully, you may see branches of the olfactory nerve that supply them. The receptive olfactory cells are modified olfactory neurons. Human conchae are neither as large nor as complexly folded as in other mammals, such as the pig. Accordingly, human beings have a poor sense of smell relative to that of most other mammals.

EXERCISE

SEVEN

Nervous Coordination: Nervous System

W E DISCUSSED THE GENERAL ASPECTS of integration in Exercise 6. In this exercise we will consider the structure of the nervous system and its role in integrating body activities. Grossly, the nervous system can be divided into the **central nervous system**, which consists of the brain and spinal cord, and the **peripheral nervous system**, which is composed of the cranial, spinal, and autonomic nerves. Most of the peripheral nerves are mixed, being made up of hundreds of long processes, called axons, of sensory and motor neurons that carry impulses from receptors to the central nervous system and from the central nervous system to the effectors, respectively. Most of the interconnections between sensory and motor neurons are made through interneurons located within the central nervous system, and this is where the integrative activities of the nervous system occur.

A. SPINAL CORD
AND SPINAL NERVES

A.1. Dissection of the Spinal Cord
and the Spinal Nerves

Most breeds of pigs have 39 pairs of spinal nerves, but the number varies according to variations in the number of thoracic and lumbar vertebrae. The nerves are named according to the vertebral region from which they arise: cervical nerves, thoracic nerves, lumbar nerves, sacral nerves, and caudal nerves. The spinal cord and the origin of the spinal nerves can be exposed easily in the fetal pig because the vertebral column in which they lie is not yet completely ossified. Remove the skin from the back of your specimen in the thoracic region and cut away enough back muscles to uncover about 8 centimeters of the vertebral column. Completely expose the spinous processes and

vertebral arches of the vertebrae (see Exercise 1). As this is done, threadlike branches of spinal nerves (Figs. 7-1 and 7-2) may be seen extending into the muscles of the back. Try to save a few of them.

Using a pair of scissors, cut off the tops of the vertebral arches to expose the **vertebral canal**, in which the spinal cord lies. The **spinal cord** and spinal nerves are surrounded by connective tissue sheaths known as **meninges**, the outermost of which is the tough **dura mater**. Notice that each spinal nerve bears an enlargement, the **spinal ganglion** (dorsal root ganglion), which is situated more or less within the **intervertebral foramen** through which the **spinal nerve** leaves the vertebral canal. The spinal ganglia will be seen better if you break away the vertical bars of bone, which form the lateral walls of the vertebral canal, between successive intervertebral foramina.

Spinal Cord and Spinal Nerves

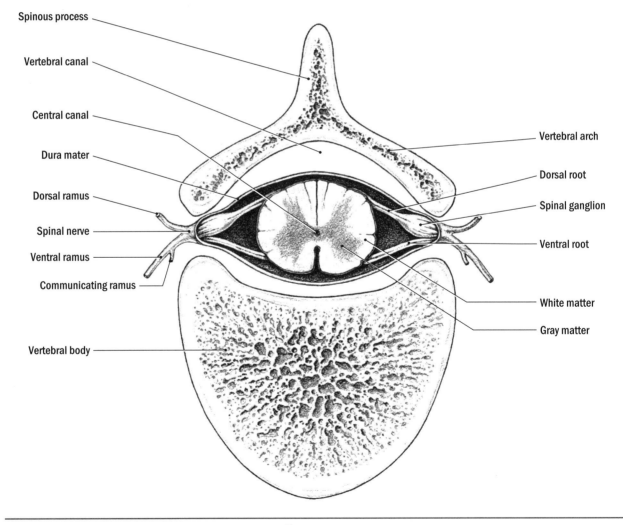

Spinous process

Vertebral canal

Central canal

Dura mater

Dorsal ramus

Spinal nerve

Ventral ramus

Communicating ramus

Vertebral body

Vertebral arch

Dorsal root

Spinal ganglion

Ventral root

White matter

Gray matter

FIGURE 7-1
Diagrammatic cross section through a thoracic vertebra and the spinal cord of a mammal.

Notice in Fig. 7-1 that the spinal nerves are formed by the union of their dorsal and ventral roots, which emerge from the spinal cord. To see the roots of the spinal nerves in your specimen, slit open the dura mater in the midline of the spinal cord with a pair of fine scissors and carefully peel it to the side. Observe that each **dorsal root** consists of a half dozen or more tiny **dorsal rootlets**, which come together at the spinal ganglion. Similarly, the **ventral root** consists of many **ventral rootlets**.

Just beyond the point at which the dorsal and ventral roots unite to form a spinal nerve, the latter divides into branches, or rami (Figs. 7-1 and 7-2). A **dorsal ramus** supplies the muscles and skin of the back, a **ventral ramus** supplies those of the flanks, and one or two **communicating rami** carry sympathetic nerve fibers to the sympathetic cord (see later). You probably will not be able to see the communicating rami without magnification.

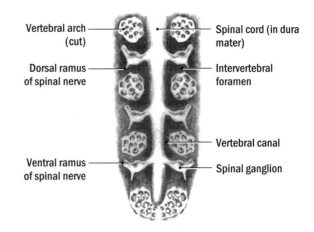

FIGURE 7-2
Dorsal view of a dissection of the spinal cord and spinal nerves of a fetal pig. The dura mater has not been cut open.

A.2. Microscopic Structure of the Spinal Cord and the Spinal Nerves

Study further aspects of the spinal cord and nerves by examining histological slide preparations of these structures from a representative vertebrate, such as a frog or a cat (Fig. 7-3). The ventral surface of the **spinal cord** can be recognized by a conspicuous **ventral fissure**. As in all chordates, the spinal cord of vertebrates is hollow and contains a small **central canal** lined by a nonnervous **ependymal epithelium**. A lymphlike cerebrospinal fluid circulates in this space. The rest of the spinal cord and the spinal nerves are composed primarily of nerve cells, or **neurons**, and their processes. They are distributed in such a way within the spinal cord that one can recognize a centrally located, somewhat butterfly-shaped **gray matter** and a peripheral **white matter** (Fig. 7-1). These terms derive from the color of these areas in fresh preparations; with certain stains, the gray matter may actually appear lighter than the white matter (Fig. 7-3).

The gray matter contains the enlarged cell bodies of neurons and their slender cell processes, which appear grayish because they are not enveloped by the whitish myelin sheath consisting mostly of lipids. The **cell bodies** of the **motor neurons** form a conspicuous group of neurons in the ventrolateral part of the gray matter (Fig. 7-3A). They are large, triangular or diamond-shaped cells with a conspicuous, clear nucleus containing a prominent, eccentric nucleolus. Nerve impulses from other neurons are received directly by these cell bodies or by short processes known as **dendrites** that attach to most of the angles of the cell bodies. Impulses then travel along long processes, the **axons**, only one of which leaves each cell body, passes through the ventral root of a spinal nerve, and is distributed to the skeletal muscles of the body (Fig. 7-3B).

The white matter consists mostly of myelinated axons of neurons that carry impulses from the spinal cord

to centers in the brain, or from the brain down to the motor neurons. These axons are seen in cross section. The few small nuclei that you may see among them belong to cells that ensheath the axons and deposit the myelin, to connective tissue cells, or to cells in the walls of blood vessels.

Sensory neurons from receptor cells enter the spinal cord through the dorsal root of a spinal nerve. Their cell bodies are located in the **spinal ganglion** (Fig. 7-3A). Some sensory neurons ascend to the brain in the white matter. Others enter the dorsal part of the gray matter, where they terminate and are relayed elsewhere by **interneurons**, whose cell bodies are difficult to distinguish in preparations of the type you are studying. Yet other sensory neurons continue through the gray matter to terminate on the cell bodies of the motor neurons. It is unlikely that you will see the communicating rami connecting the ganglion and spinal nerve. The terminations of sensory neurons, either directly on motor neurons or indirectly through interneurons, form **reflex arcs** that are responsible for simple **spinal reflexes** (Fig. 7-3B).

In some histological slides, a **sympathetic ganglion** may be seen attached to the ventral surface of a spinal nerve a short distance distal to the union of the dorsal and ventral roots of the spinal nerve (see the following section).

A.3. Sympathetic Cord and Autonomic Nervous System

The cervical extension of the sympathetic cord was seen earlier lying deep to the vagus nerve between the common carotid artery and internal jugular vein (see Exercise 4). Turn your pig on its back and spread open its left side

Spinal Cord and Spinal Nerves

101

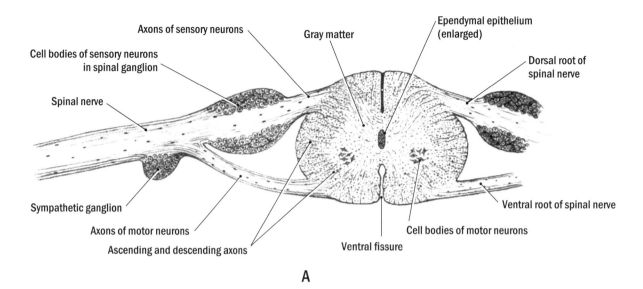

Axons of sensory neurons

Cell bodies of sensory neurons
in spinal ganglion

Gray matter

Ependymal epithelium
(enlarged)

Dorsal root of
spinal nerve

Spinal nerve

Sympathetic ganglion

Axons of motor neurons

Ascending and descending axons

Ventral fissure

Cell bodies of motor neurons

Ventral root of spinal nerve

A

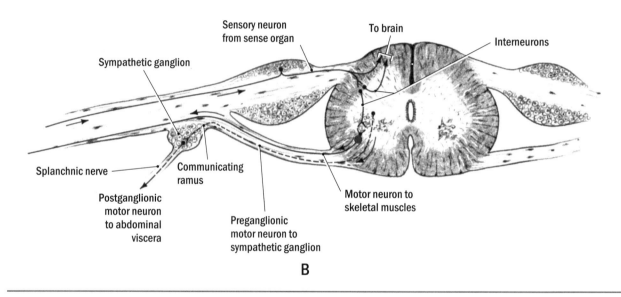

Sensory neuron
from sense organ

To brain

Interneurons

Sympathetic ganglion

Splanchnic nerve

Communicating
ramus

Postganglionic
motor neuron
to abdominal
viscera

Preganglionic
motor neuron to
sympathetic ganglion

Motor neuron to
skeletal muscles

B

FIGURE 7-3
Spinal cord and spinal nerve of a frog.
(A) Drawing of a transverse section.
(B) Diagram of the major types of neurons and their interconnections.

of the thorax as wide as you can. Pull the heart and lungs ventrally and toward the right side of the specimen, and bend the thoracic wall dorsally. A slender nerve running longitudinally against the thoracic wall lies slightly dorsal and lateral to the aorta and left azygos vein (see Fig. 4-5). This is the thoracic part of the **sympathetic cord**; the segmental swellings on it are **sympathetic ganglia**. The communicating rami of the spinal nerves connect with these ganglia. A corresponding sympathetic cord lies on the right side of the body.

The sympathetic cords can be traced caudally into the abdomen, where they lie dorsal to the aorta or caudal

vena cava. At about the level of the diaphragm, several threadlike **splanchnic nerves** extend from each sympathetic cord to **collateral sympathetic ganglia,** which surround the bases of the coeliac and cranial mesenteric arteries. Most of these structures were probably picked away when the arteries were dissected (Exercise 4). Delicate sympathetic nerve fibers, which cannot be seen without high magnification, lead from the collateral sympathetic ganglia along the arteries to the abdominal viscera. Organs in the head and neck receive sympathetic nerve fibers by means of the extension of the sympathetic cords into the neck.

The sympathetic cords are a part of the autonomic nervous system. This system consists of special motor neurons that supply most of the glands of the body and the involuntary muscles of the heart, blood vessels, urinary bladder, and gut wall. It is a unique feature of the autonomic nervous system that there is always a relay in a peripheral ganglion (Fig. 7-3B). That is, the motor neuron that leaves the central nervous system (**preganglionic neuron**) synapses with a second motor neuron (**postganglionic neuron**) whose cell body lies within a peripheral ganglion. The postganglionic neuron then continues to the effector organ.

The sympathetic nervous system and the less conspicuous parasympathetic nervous system together form the **autonomic nervous system**. Most organs supplied by the autonomic nervous system receive nerve fibers from both parts, and these nerve fibers have antagonistic effects on the organs. In general, sympathetic stimulation, together with secretions of the medulla of the adrenal gland, help the body adjust to stress by providing more energy to skeletal muscles (e.g., blood pressure increases, release of glucose into the blood). This response is the "flight or fight" reaction. Parasympathetic stimulation promotes the conservation and restoration of energy (e.g., storage of glucose); it is a "rest and digest" reaction.

Sympathetic nerve fibers leave the central nervous system through the thoracic nerves and certain lumbar nerves; they enter the sympathetic cord, and from there they are distributed to the organs they supply. The peripheral relay between preganglionic and postganglionic fibers occurs in the sympathetic ganglia on the sympathetic cord or in the collateral sympathetic ganglia. Cell bodies of the postganglionic sympathetic neurons can be seen microscopically if a sympathetic ganglion is present on the histological slide of the spinal cord and nerve (Fig. 7-3A).

Parasympathetic nerve fibers leave the central nervous system through certain cranial nerves and through sacral spinal nerves. The vagus nerve (cranial nerve X), dissected in Exercise 4, contains preganglionic parasympathetic nerve fibers that supply thoracic and abdominal viscera. Preganglionic parasympathetic nerve fibers to the eye and glands in the head leave through cranial nerves III, VII, and IX. The peripheral relay of parasympathetic nerve fibers is in or very near the organ supplied, such as in intramural ganglia (see Exercise 3). Thus the postganglionic nerve fibers are very short.

B. BRAIN AND CRANIAL NERVES

The structure of the mammalian brain can be better studied on a sagittal section of a sheep brain than on the brain of the fetal pig. But if a sheep's brain is not available, the pig brain can be removed by cutting away the skin from the head and carefully picking away the bones from the top and sides of the skull (Fig. 7-4). The tough membrane (dura mater) covering the brain should be left intact as long as possible. The following description refers to the sheep brain but can be applied to the brain of many other mammals, including pigs and human beings.

B.1. Meninges

The mammalian brain is surrounded by several layers of connective tissue, or **meninges**. A vascular **pia mater** closely covers the surface of the brain and follows all of its convolutions. The more delicate **arachnoid** lies superficial to the pia mater and can usually be separated from the pia mater where it crosses, but does not dip into the indentations on the brain surface. In life, a tough, protective **dura mater** covers the other meninges. If you are studying a preserved sheep brain, the dura mater has probably been removed from your specimen, so it should be looked at on demonstration preparations.

A lymphlike **cerebrospinal fluid**, which is secreted by certain vascularized membranes into the cavities of the brain, circulates within the central nervous system and between the pia mater and the arachnoid. It forms a liquid hydrostatic cushion around the brain and spinal cord, insulating these delicate structures from the surrounding bone of the skull and vertebrae and protecting them from mechanical injury. It also makes an important contribution to brain nutrition and waste removal.

B.2. Endocrine Glands Attached to the Brain

Two important endocrine glands are associated with the brain. The **pineal gland** can be seen in a sagittal section of the sheep brain. It lies beneath the caudal end of the large cerebral hemisphere and is attached to the epithalamus (see later and Fig. 7-8). The pineal gland of mammals has evolved from a light-receptive third eye found on the top of the head of some reptiles and ancestral vertebrates. Indeed, its secretory cells are modified photoreceptors. These cells synthesize and release the hormone **melatonin**, particularly under dark conditions. Melatonin synthesis is inhibited by light. The seasonal enlargement of the ovary and testis in rodents and some other mammals that breed in the spring results from the reduction of melatonin production as the length of the day increases. The pineal gland may also have a role in regulating other seasonal and diurnal cycles. It has been proposed that human jet lag happens because the diurnal synthesis of melatonin gets out of phase with the normal dark–light cycle.

The **hypophysis**, or pituitary gland, is a dark, ovular gland attached to the ventral surface of the brain by a narrow stalk known as the **infundibulum** (Fig. 7-7 and 7-8). One part of the hypophysis develops embryonically as

Brain and Cranial Nerves

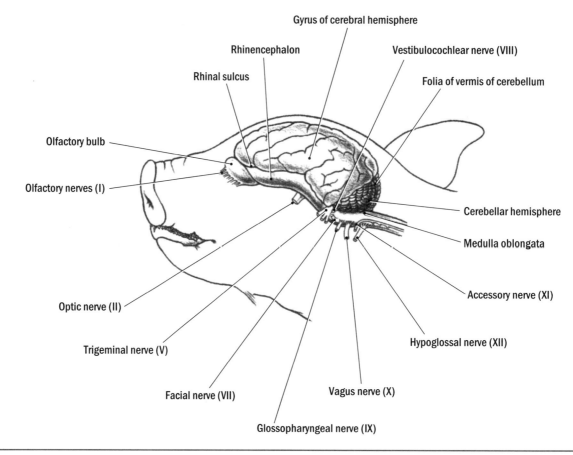

FIGURE 7-4
Lateral view of the brain and cranial nerves of a fetal pig.

Labels in figure:
- Gyrus of cerebral hemisphere
- Rhinencephalon
- Rhinal sulcus
- Vestibulocochlear nerve (VIII)
- Folia of vermis of cerebellum
- Olfactory bulb
- Olfactory nerves (I)
- Optic nerve (II)
- Trigeminal nerve (V)
- Facial nerve (VII)
- Glossopharyngeal nerve (IX)
- Vagus nerve (X)
- Hypoglossal nerve (XII)
- Accessory nerve (XI)
- Medulla oblongata
- Cerebellar hemisphere

Brain and Cranial Nerves

an outgrowth from the floor of the brain (diencephalon) and the other part as an ingrowth from the roof of the mouth.

The hypophysis produces, or stores, many hormones that have wide-ranging effects. **Melanophore stimulating hormone** promotes the synthesis and dispersion of the dark pigment melanin. **Thyrotropic hormone** stimulates the thyroid gland. **Adrenocorticotropic hormone** increases the activity of the adrenal cortex (see Exercise 4). **Growth hormone** promotes body growth. Two **gonadotropic hormones** (**follicle stimulating hormone** and **luteinizing hormone**) regulate the activity of the ovary and testis (see Exercise 5). **Prolactin** stimulates the synthesis of milk by the mammary glands. **Oxytocin** helps release milk during lactation. **Antidiuretic hormone** promotes the reabsorption of water from parts of the kidney tubules. The last two hormones are actually synthesized in a part of the brain (the hypothalamus, see later) but are stored and released by the hypophysis.

B.3. Brain Structure

The different regions of the adult brain develop from five distinct swellings of the early embryonic brain, so it is convenient to divide the adult brain into regions having the same names as the original embryonic swellings. From anterior to posterior, these regions are (1) telencephalon, (2) diencephalon, (3) mesencephalon, (4) metencephalon, and (5) myelencephalon. The extent of these regions and the major structures that develop from them are shown in Fig. 7-5.

Major parts of the brain can be seen in surface views (Figs. 7-6 and 7-7), in a sagittal section (Fig. 7-8), and, if demonstration material is available, in one or more transverse sections (Fig. 7-9). The most conspicuous parts of the brain are the large, paired **cerebral hemispheres**, which constitute most of the **telencephalon**. They are the primary integration centers of the mammalian brain. The dorsolateral part of each hemisphere is convoluted, forming many ridges, or **gyri**, that are separated by grooves, or

104

sulci. This part of the cerebral hemispheres has evolved only in mammals and is known as the **neopallium**. Make a tangential slice across a part of the neopallium, or examine a transverse section through the brain. Notice that the surface forms a gray **cerebral cortex** and that **white matter** lies beneath this. As in the spinal cord, **gray matter** is composed largely of the cell bodies of neurons and unmyelinated fibers and white matter is composed largely of myelinated axons of neurons. The convolutions of the cerebral surface greatly increase the area available for nerve cell bodies.

The axons in the white matter belong to many groups of interneurons that cannot be distinguished grossly, but you should know what they represent collectively. **Sensory fibers** project from lower centers in the brain and bring sensory information to specific cerebral regions. **Association fibers**, which interconnect different gyri of the same hemisphere, and **commissural fibers**, which interconnect the two hemispheres, make it possible for different kinds of sensory information to be integrated and compared with memories of past experience. Eventually, motor impulses are initiated in specific motor regions of the cerebral cortex and sent out on **motor fibers** to centers in the brain stem and to the spinal cord.

The cerebral hemispheres of human beings are much larger and more complexly folded than those of sheep.

A conspicuous **rhinal sulcus** (Figs. 7-6 and 7-7) separates the dorsolateral neopallium from a rather large, more ventrally situated part of the cerebral hemisphere, the **rhinencephalon**. From an evolutionary viewpoint, the rhinencephalon is one of the oldest parts of the telencephalon. In ancestral vertebrates, it integrated olfactory (smell) impulses, and it remains a primary olfactory center in all vertebrates. An **olfactory bulb** forms the rostral part of the rhinencephalon and receives branches of the olfactory nerve returning from the nasal cavities. A **lateral olfactory tract** extends from the olfactory bulb to caudal parts of the rhinencephalon.

Several bundles of commissural fibers that interconnect the two sides of the telencephalon can be seen in a sagittal section (Fig. 7-8) and in a transverse section (Fig. 7-9). The largest and most dorsal of these commissures is the **corpus callosum**, which connects the neopallium of each side of the brain. The wall of the hemisphere ventral to the corpus callosum is a thin **septum pellucidum**. Break through this septum and notice part of a large chamber, the **lateral ventricle**, within one of the hemispheres. There is one lateral ventricle in each hemisphere, and they constitute the first and second ventricles.

Another band of white fibers, the **fornix**, lies ventral to the septum pellucidum. It curves rostrally and ventrally toward the hypothalamus, or floor of the diencephalon, but soon disappears from the sagittal plane.

The fornix is part of a pathway that interconnects olfactory centers in the telencephalon with the hypothalamus and thalamus. This pathway and associated parts of the brain form the **limbic system**. The limbic system influences many aspects of motivational and emotional behavior related to self-preservation and species preservation, including feeding, drinking, fighting, and reproduction. The limbic system also participates in the formation of short-term memories.

Near the point where the rostral end of the fornix disappears from the sagittal plane is a small round bundle of fibers, the **anterior commissure**, which connects the major olfactory parts of the two hemispheres. A thin vertical partition extends from the anterior commissure to the crossing of the **optic nerves**, the **optic chiasma**, on the ventral surface of the brain. This is the **lamina terminalis**, the most rostral part of the embryonic brain in the midline. The cerebral hemispheres develop as rostrolateral expansions from this region. The optic fibers then continue in optic tracts across the lateral surface of the diencephalon. Most optic fibers terminate in the thalamus, but some continue to the superior colliculi (see later).

The cerebral hemispheres of fishes, amphibians, and reptiles are relatively small and are situated entirely rostral to the other regions of the brain. As these hemispheres assumed additional functions in the course of evolution toward mammals, they enlarged and grew caudally over the dorsal surface of much of the brain. The overgrowth that has occurred is evident in the sagittal section through the brain.

The major brain region caudal to the telencephalon, which is largely covered by the cerebral hemispheres, is the **diencephalon** (Fig. 7-5). It extends from the fornix, anterior commissure, and lamina terminalis caudally to include a conspicuous bump on the brain surface, the pineal gland, which you have already seen. The diencephalon contains a narrow chamber, the **third ventricle**, which surrounds the interthalamic adhesion (see later) and whose extent can be recognized by its shiny lining of ependymal epithelium (Fig. 7-8). (Further dissection may be necessary to expose the ventricle.)

The third ventricle is crossed by a large circular mass of nervous tissue, the **interthalamic adhesion**. Inconspicuous **interventricular foramina** connect the two lateral ventricles with the third ventricle. Vascular tissue forms much of the roof of the diencephalon, and tufts from it dip into the third ventricle. These tufts form one of the **choroid plexuses**, which are vascular networks that secrete cerebrospinal fluid into the ventricles (Fig. 7-9).

The diencephalon can be divided into the **epithalamus**, lying dorsal to the third ventricle; the **hypothalamus**, ventral to it; and the halves of the **thalamus** on each side of the third ventricle. The epithalamus consists of the vascular roof of the ventricle, the pineal gland, and adjacent nervous tissue related to olfaction.

*Brain and
Cranial Nerves*

105

The hypothalamus is the conspicuous oval area that can be seen on the ventral surface of the brain caudal to the optic chiasma (Fig. 7-7). The hypophysis, which you have already seen, attaches to the hypothalamus; if it has already been removed, you will see only its stalk, which is called the **infundibulum**. The hypothalamus integrates many autonomic functions: sleep, body temperature, water balance, appetite, and both carbohydrate and fat metabolisms. It exerts its influence by means of motor pathways that go out from it, and by producing **releasing** and **inhibiting hormones** that flow through small blood vessels leading to the hypophysis and affect the activity of the hypophysis.

The thalamus includes the interthalamic adhesion and all of the lateral wall of the diencephalon. Its lateral part, which is covered by the cerebral hemisphere, can be exposed by carefully pulling the caudal end of the cerebral hemisphere laterally. (Do not tear the brain.) Its thickness can be seen in a transverse section (Fig. 7-9). As the cerebrum has increased in importance during the evolu-

Brain and Cranial Nerves

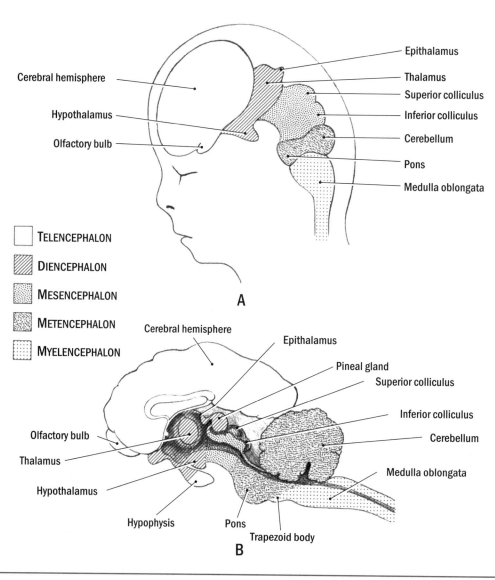

□ TELENCEPHALON

▨ DIENCEPHALON

▦ MESENCEPHALON

▨ METENCEPHALON

▦ MYELENCEPHALON

FIGURE 7-5
Brain regions.
(A) Lateral view of an 8-week-old human embryo.
(B) Diagram of a sagittal section of an adult sheep brain.

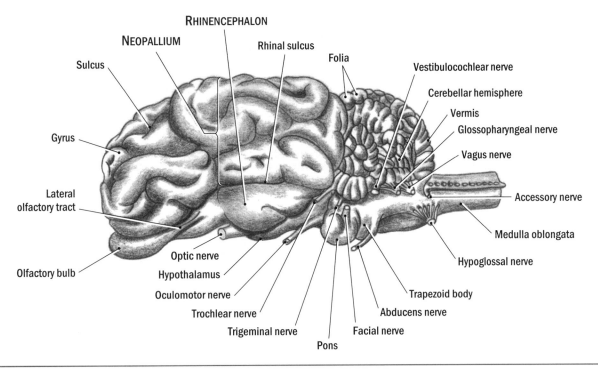

FIGURE 7-6
Lateral view of a sheep brain with the stumps of the cranial nerves.

*Brain and
Cranial Nerves*

tion of mammals, so has the thalamus, which interconnects the cerebrum with many other parts of the brain. All sensory information, apart from olfaction, entering the spinal cord or brain on primary sensory neurons proceeds to the thalamus on interneurons (Fig. 7-10). Most of these interneurons decussate, or cross to the opposite side of the central nervous system as they ascend to the brain. Sensory information is sorted in the thalamus, and some or all of it is relayed on other interneurons that contribute to a bundle, the **internal capsule** (Figs. 7-9 and 7-10), which passes through the base of the cerebrum to specific sensory areas of the cerebral cortex. Some involuntary motor activity initiated in the cortex is also relayed in the thalamus on its way to motor centers in the brain stem. But the thalamus is more than a switchboard. Although many of its functions are as yet undiscovered, there is much evidence that it interacts with the cerebral cortex in many of the higher mental processes.

The third major region of the brain, the **mesencephalon**, lies between the diencephalon and the large cerebellum (Fig. 7-5). It, too, is largely covered by the cerebrum; hence, it can best be seen in a sagittal section. Dorsally, the mesencephalon bears two pairs of round swellings; each of the larger and more rostral ones is a **superior colliculus**, and each of the others is an **inferior colliculus** (Fig. 7-8). The superior colliculi, also known as the optic lobes, are the major integration centers in the brain of fishes and amphibians, but in mammals most of their integrative function has been transferred to the cerebrum. However, some fibers of the optic nerve still end in the superior colliculi, and they retain their function as centers for certain eye reflexes, including the pupillary reflex, accommodation, and eyeball movements. Certain auditory reflexes occur in the inferior colliculi.

The **trochlear nerve** to one of the extrinsic ocular muscles leaves the brain just caudal to the inferior colliculus. Ventrolaterally, the mesencephalon bears a pair of large fiber tracts, the **cerebral peduncles** (Figs. 7-7 and 7-8). Fibers of the most important voluntary motor pathway, known as the **pyramidal tract**, passes directly through the internal capsule, a fiber pathway through the thalamus, and the cerebral peduncles on their way from motor cell bodies in the cerebral cortex to the cell bodies of motor neurons of certain cranial nerves and the spinal nerves (Fig. 7-10). Like the ascending sensory fibers, most of the descending pyramidal fibers decussate before reaching the motor neurons. **Oculomotor nerves** to most extrinsic ocular muscles attach to the cerebral peduncles. The **cerebral aqueduct** runs through the center of the mesencephalon and connects the third ventricle with the **107**

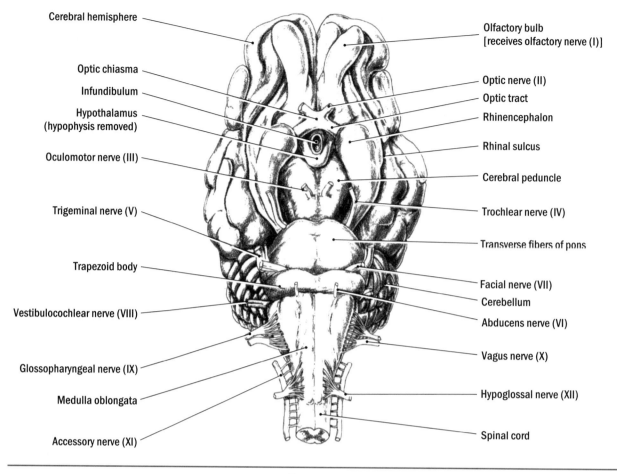

Cerebral hemisphere

Optic chiasma

Infundibulum

Hypothalamus
(hypophysis removed)

Oculomotor nerve (III)

Trigeminal nerve (V)

Trapezoid body

Vestibulocochlear nerve (VIII)

Glossopharyngeal nerve (IX)

Medulla oblongata

Accessory nerve (XI)

Olfactory bulb
[receives olfactory nerve (I)]

Optic nerve (II)

Optic tract

Rhinencephalon

Rhinal sulcus

Cerebral peduncle

Trochlear nerve (IV)

Transverse fibers of pons

Facial nerve (VII)

Cerebellum

Abducens nerve (VI)

Vagus nerve (X)

Hypoglossal nerve (XII)

Spinal cord

*Brain and
Cranial Nerves*

FIGURE 7-7
Ventral view of a sheep brain, with the stumps of the cranial nerves.

large **fourth ventricle** located in the caudal parts of the brain (Fig. 7-8).

The large **cerebellum** and the ventral part of the brain to which it attaches constitute the fourth major brain region, the **metencephalon** (Fig. 7-5). Just as the cerebrum became enlarged and gained more control over the activity of the body during the evolution of mammals, the cerebellum, too, has enlarged and has become interconnected with motor areas in the cerebrum. The cerebellum resembles the cerebrum in being composed of many folds, the **folia**, and in having a gray cortex. Its white matter forms a treelike pattern called the **arbor vitae** (Fig. 7-8).

The cerebellum can be divided into several parts, the most conspicuous of which are a median **vermis** and a pair of lateral **cerebellar hemispheres** (Fig. 7-6). The primary sensory inflow of the cerebellum is from the part of the inner ear that detects changes in body position and

from the proprioceptive organs in muscles, which indicate their current state of contraction. It also receives copies, so to speak, of motor directives from the cerebrum (Fig. 7-10). These terminate in the ventral part of the metencephalon, the **pons**. Neurons receiving these impulses decussate in the transverse fibers of the pons and lead to the cerebellar cortex. The cerebellum is, therefore, a crucial center for muscular coordination, for monitoring the orientation and motor activity of the body, and for initiating corrective impulses that either go back to the cerebrum by way of the thalamus or go directly to the cell bodies of the motor neurons. It adjusts body and eye movements to changed positions of the head, regulates postural and locomotor movements relative to current positions and activities of the limbs, and helps regulate the timing and duration of muscle contractions, especially those related to skilled, voluntary movements. Several cranial nerves attach just caudal to the pons, and

108

their stumps can be found on good specimens (see Figs. 7-6 and 7-7): the **trigeminal, abducens, facial**, and **vestibulocochlear nerves**.

The fifth and final region of the brain is the **myelencephalon** (Fig. 7-5). It consists of the **medulla oblongata**, which merges caudally with the spinal cord. At the front of the underside of the medulla oblongata, just caudal to the pons, there is a small transverse band of fibers, the **trapezoid body** (Fig. 7-7). This is an acoustic commissure. The pyramidal motor system forms a pair of bulges, the **pyramids**, on the ventral surface of the medulla oblongata caudal to the trapezoid body.

The **fourth ventricle** continues from the metencephalon into the medulla oblongata and has a very thin, vascular roof that forms another choroid plexus. Microscopic perforations in the roof permit cerebrospinal fluid to escape from the fourth ventricle and circulate between the pia mater and arachnoid. The fluid is ultimately reabsorbed from certain parts of the meninges into the venous system. Many visceral activities are regulated by centers in the medulla oblongata: rate of heart beat, blood pressure, breathing movements, salivation, swallowing. Stumps of the remaining cranial nerves can be found attached to the ventral surface of the medulla oblongata on good specimens: **glossopharyngeal, vagus, accessory**, and **hypoglossal nerves**.

B.4. Cranial Nerves

The stumps of many of the cranial nerves have been found during your study of the brain. Look for them again (Figs. 7-6 and 7-7). Many are very fragile, and it is unlikely that all can be seen on a single specimen. Mammals have twelve pairs of cranial nerves and each is given a number, according to its position in human beings, as well as a name.

I. Olfactory Nerve   This nerve consists of groups of processes of olfactory cells entering the olfactory bulb from the nasal cavity. These processes are usually torn off when the brain is removed from the skull.

II. Optic Nerve   The sensory nerve from the retina, whose fibers terminate in the thalamus and superior colliculus.

Brain and Cranial Nerves

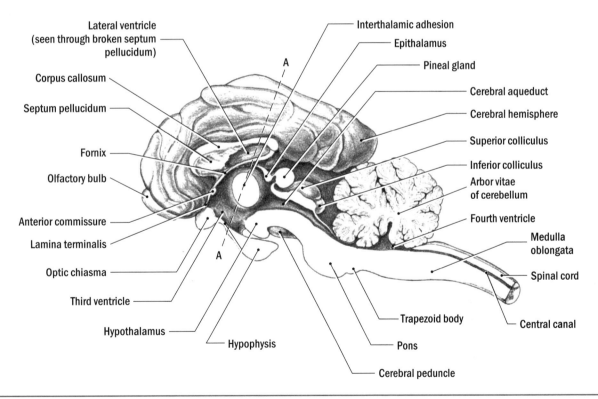

FIGURE 7-8
Dissection of a sagittal section of a sheep brain. See Fig. 3-3 for a sagittal section of the brain of a fetal pig.
The transverse line (A–A) is in the plane of the transverse section shown in Fig. 7-9.

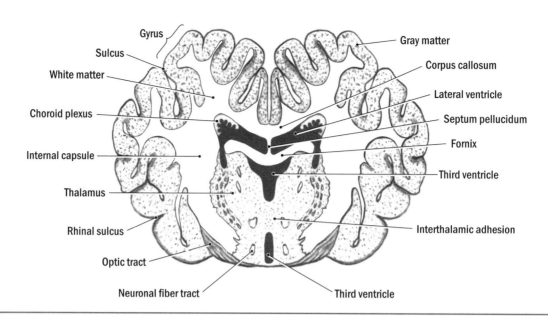

FIGURE 7-9
Transverse section of the sheep brain made slightly caudal to the optic chiasma in the plane A–A shown in Fig. 7-8.

Brain and Cranial Nerves

Because the retina develops embryonically as an outgrowth of the brain, the optic nerve is technically a brain fiber tract and not a typical peripheral nerve. One-half of the fibers in each optic nerve (those from the caudal or lateral part of the retina) terminate on the side of the brain they entered; the other half cross to the opposite side of the brain, and this crossing forms the optic chiasma. Thus, images from each eye are projected to each side of the brain. This dual projection plays a role in stereoscopic vision and depth perception.

III. Oculomotor Nerve Primarily a motor nerve to four of the seven small, straplike muscles that move the eyeball (extrinsic ocular muscles) and to a muscle that moves the upper eyelid. It also carries parasympathetic motor fibers to muscles in the ciliary body involved in accommodation and to those in the pupil. Like most motor nerves, it carries a few sensory proprioceptive nerve fibers returning from the muscles and providing information on the degree and extent of muscle contraction.

IV. Trochlear Nerve A motor nerve to one of the extrinsic ocular muscles.

V. Trigeminal Nerve A mixed nerve carrying both sensory and motor nerve fibers. It is the primary cutaneous sensory nerve of the head, but it also carries motor nerve fibers to the jaw muscles. Jaw muscles are bran-

chiomeric muscles, for they evolved from the first visceral (branchial) arch muscles of ancestral fishes (see Exercise 2).

VI. Abducens Nerve The last of the motor nerves to the extrinsic ocular muscles; supplies one muscle.

VII. Facial Nerve A mixed nerve. It is the motor nerve of facial muscles (see Exercise 2), and it supplies parasympathetic nerve fibers to salivary and tear glands. Its sensory fibers arise from taste buds on the anterior two-thirds of the tongue. The muscles supplied by this nerve are branchiomeric muscles, having evolved from the second visceral arch muscles of ancestral fishes (see Exercise 2).

VIII. Vestibulocochlear Nerve (acoustic nerve) The sensory nerve from the ear. Its sensory fibers carry both impulses related to equilibrium from the vestibular parts of the ear and auditory impulses from the cochlea—hence, the name for the nerve.

IX. Glossopharyngeal Nerve A mixed nerve. It is the motor nerve to pharyngeal muscles, which are branchiomeric muscles associated with the third visceral arch. It also carries parasympathetic nerve fibers to salivary glands. Its sensory fibers arise from taste buds on the posterior third of the tongue and from part of the lining of the pharynx.

X. Vagus Nerve A mixed nerve. Its motor nerve fibers supply part of the pharynx, larynx, heart, stomach, and intestinal region. The pharyngeal and laryngeal muscles are branchiomeric muscles, having evolved from muscles of caudal visceral arches in ancestral fishes. The motor nerve fibers that go to the heart and gut belong to the parasympathetic part of the autonomic nervous system, but the sensory nerve fibers returning impulses from many internal organs (larynx, lungs, heart, stomach) are not a part of the autonomic nervous system. The glossopharyngeal and vagus nerves connect to the brain by many small rootlets that lie close together; the nerves separate from each other after they emerge from the skull.

XI. Accessory Nerve A motor nerve to certain branchiomeric muscles of the neck and shoulder (sternomastoid, cleidomastoid, see Exercise 2). It also contains the usual proprioceptive nerve fibers.

XII. Hypoglossal Nerve A motor nerve to the muscles of the tongue.

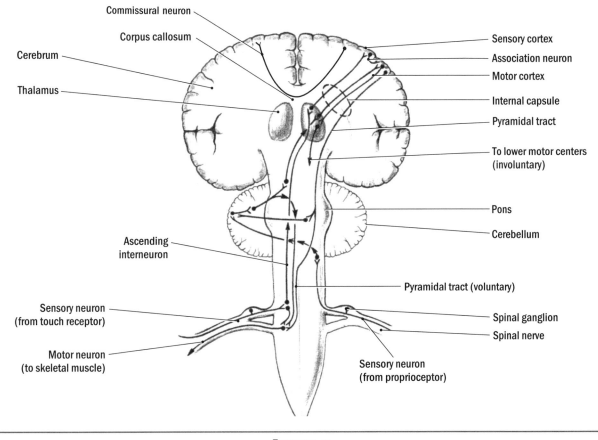

Brain and Cranial Nerves

FIGURE 7-10
Wiring diagram in a ventral view of the nervous system, showing the interrelations of neurons that form some of the major sensory and motor pathways. Each dark, round spot represents the cell body of a neuron, and lines extending from them represent axons. Impulses travel from the cell bodies to the ends of the axons.

GLOSSARY

Glossary of Vertebrate Anatomical Terms

T HIS GLOSSARY IS A BASIC VOCABULARY of anatomical terms that students will encounter in using *Anatomy and Dissection of the Fetal Pig, Anatomy and Dissection of the Rat, Dissection of the Frog,* and many other exercises dealing with vertebrates. Most entries include the term, its pronunciation, its derivation, and its definition.

The pronunciation is given by a simple phonetic spelling in brackets. Stressed syllables are marked by a prime (ˊ); others are separated by hyphens. Long vowels are indicated by the macron (¯); short ones, by the breve (˘).

The classical derivation in parentheses typically includes an abbreviation of the language of the word, the original word in italics, and the meaning of the original word. Usually only the nominative singular is given for Greek and Latin nouns, but the genitive (gen.) sometimes is used if it is closer to the root. Latin adjectives are given only in the masculine form of the nominative singular. Greek and Latin verbs usually are given in the first person present tense because this is closer to the root than the infinitive (but the meaning is given in the infinitive). The present or past participle (pres. p. or p. p.) is given when one of them is closer to the root. Many Greek and Latin words have a combining form that is used when the word is used in combination with other words. Combining forms are indicated by a hyphen before or after the word, for example, L. **Inter-** = between, as in interstitial.

When the English and classical terms are identical, the term is not repeated in the derivation, for example, **Acetabulum** (L. = vinegar cup). When two or more successive terms use the same root, the derivation of the root is given only for the first term. The origin of many repetitive terms is given only under the first entry. For example, **Ligamentum arteriosum** is defined the way this combination of terms is used, but the derivations of ligamentum and arteriosum will be found under **Ligament** and **Artery**. Names of individual muscles that are descriptive of an easily recognized feature of the muscle (its shape or attachments) are not given, but the component parts of less familiar ones are included. For example, the omohyoid muscle is not listed, but **Omo-** and **Hyoid** are. Similarly, names of blood vessels and nerves that simply state the organ supplied are omitted, but the organs are usually listed.

This glossary is not exhaustive. The pronunciation, derivation, and meaning of additional terms can be found in unabridged and medical dictionaries.

Abdomen [ab′dō-men] (L. from *abdo* = to conceal).
The part of the body containing the visceral organs, limited in mammals to the part caudal to the diaphragm.

Abducens nerve [ab-du′senz] (L. leading away, from *ab-* = away from + *duco*, pres. p. *ducens* = leading).
The sixth cranial nerve; carries motor fibers to an extrinsic ocular muscle.

Abduction [ab-dŭk′shŭn] (L. *duco*, p. p. *ductus* = to lead).
Muscle action that carries a body part away from a point of reference, often the midventral line of the body.

Accessory nerve.
The eleventh cranial nerve of amniotes. It carries motor fibers to certain branchiomeric shoulder muscles: the trapezius and sternocleidomastoid muscles.

Acetabulum [as-ĕ-tab′yū-lŭm] (L. = vinegar cup, from *acetum* = vinegar).
The cup-shaped socket in the pelvic girdle that receives the head of the femur.

Achilles tendon.
See **Tendon of Achilles.**

Acromion [ă-krō′mē-on] (Gr. *akron* = tip + *omos* = shoulder).
A process on the scapula with which the clavicle articulates in species with a well-developed clavicle.

Adduction [ă-dŭk′shŭn] (L. *ad-* = toward + *duco*, p.p. *ductus* = to lead).
Muscle action that pulls a body part toward a point of reference, often the midventral line of the body.

Adrenal gland [ă-drē′năl] (L. *ad-* = toward, beside + *ren* = kidney).
An endocrine gland located cranial to the kidney (mammals) or on its ventral surface (frogs). Its medullary hormone helps the sympathetic nervous system adjust the body to stress; its cortical hormones help regulate sexual development and the metabolism of minerals, carbohydrates, and proteins. It is also called the *suprarenal gland.*

Allantois [ă-lan′tō-is] (Gr. *allos* = sausage + *eidos* = appearance).
An extraembryonic membrane of reptiles, birds, and mammals. It develops as an outgrowth of the embryonic hindgut. It accumulates waste products. Its vascularized wall serves as a gas exchange organ in embryonic reptiles and birds, and it contributes to the formation of the placenta in eutherian mammals. Its base develops into the urinary bladder of adult mammals.

Alveolus [al-vē′ō-lŭs] (L. = small cavity).
One of a group of small, thin-walled, and vascularized sacs at the termination of the mammalian respiratory tree where gas exchange occurs.

Amnion [am′nē-on] (Gr. = the membrane around the fetus).
The innermost of the extraembryonic membranes. It envelops the fetus and contains amniotic fluid.

Amniote [am′nē-ōt].
A vertebrate whose embryo has an amnion; a reptile, bird, or mammal.

Amphibian [am-fi′bē-ăn] (Gr. *amphi* = both, double + *bios* = life).
A frog, salamander, or other member of the ancestral class of terrestrial vertebrates. Usually it is aquatic as a larva and terrestrial as an adult.

Ampullary gland [am-pul′lăry] (L. *ampulla* = small vessel).
A small gland associated with the terminal end of the ductus deferens in male rats. It contributes to the seminal fluid.

Anamniote [an-am′nē-ōt] (Gr. *an* = without + *amnion* = the membrane around the fetus).
A vertebrate without an amnion; a fish or amphibian.

Antebrachium [an-te-brā′kē-ŭm] (L. *ante* = before + *brachium* = upper arm).
The forearm.

Anterior.
A direction toward the front or belly surface of a human being; sometimes also used for the head end of a quadruped, but *cranial* is a more appropriate term.

Anterior chamber.
The space within the eyeball between the cornea and iris. It is filled with aqueous humor.

Anterior commissure.
An olfactory commissure within the cerebrum. It is located just rostral to the third ventricle.

Antrum [an′trŭm] (Gr. *antron* = a cave).
An enclosed cavity within an organ, such as the antrum in a secondary follicle in the mammalian ovary.

Anura [an-yūr′a] (Gr. *an* = without + *oura* = tail).
The amphibian order to which frogs and toads belong.

Anus [ā′nŭs] (L. = the seat, anus).
The caudal opening of the digestive tract in a mammal.

Aorta [ā-ōr′tă] (Gr. *aorte* = great artery).
The major artery carrying blood from the heart to the body. It is sometimes called the dorsal aorta to distinguish it from the ventral aorta of a fish, which carries blood from the heart to the gills.

Aortic valve.
A set of three semilunar-shaped folds at the base of the mammalian aorta. It prevents a backflow of blood into the left ventricle.

Aponeurosis [ap-ō-nū-rō′sis] (Gr. *apo* = from + *neuron* = sinew, nerve).

A sheetlike tendon of a muscle.

Appendix [ă-pen′diks] (L. *appendo* = to hang something on).

A dangling extension of an organ, such as the vermiform appcndix at the end of the caecum of some mammals.

Aqueduct of Sylvius (*Franciscus Sylvius,* Dutch Anatomist, 1614–1672).

See **Cerebral aqueduct.**

Aqueous humor [ā′kwē-ŭs hyū′mer] (L. *aqua* = water + *humor* = fluid).

A watery liquid in the anterior and posterior chambers of the eyeball. It is secreted by the ciliary body.

Arachnoid [ă-rak′noyd] (Gr. *arachne* = spider + *eidos* = appearance).

One of three meninges of mammals. It is located between the dura mater and pia mater, and is connected to the pia mater by many strands, which give it a "spider web" appearance.

Arbor vitae [ar′bōr vīt′ē] (L. *arbor* = tree + *vita* = life).

The tree-shaped configuration of white matter within the mammalian cerebellum.

Archinephric duct [ar′ki-nef-rik] (Gr. *arche* = origin, beginning + *nephros* = kidney).

The duct, common to many fishes and amphibians, that drains the kidney. It also transports sperm in the male of many vertebrates. It becomes the ductus deferens in a male mammal. It is also called the *mesonephric duct.*

Areola [ă-rē′ō-lă] (L. = small space).

A small space or area, such as the small, round protuberances on the surface of a pig's chorion.

Arrector pili [ă-rek′tōr pi′li] (L. the raisor, from *arrectus* = upright + *pili* = hairs).

One of the small muscles in the skin of mammals that attach onto the hair follicles and raise the hairs.

Artery [ar′ter-ē] (L. *arteria* = artery).

A vessel that carries blood away from the heart. The blood is usually rich in oxygen, but it may be low in oxygen, as, for example, in the pulmonary arteries, which carry blood from the heart to the lungs.

Artiodactyla [ar′ti-ō-dak-til-a] (Gr. *artios* = even + *daktylos* = finger or toe).

The mammalian order that includes hoofed mammals with an even number of toes: pigs, deer, cattle.

Atlas [at′las] (Gr. mythology, a god who supported the Earth upon his shoulders).

The first cervical vertebra that supports the skull.

Atrioventricular valve [ā′trē-ō-ven-trik′yū-lar] (L. *atrium* = entrance hall + *ventriculus* = a small belly).

The valve between an atrium and ventricle of the heart. It prevents the backflow of blood from the ventricle into the atrium. The right atrioventricular valve of mammals has three cusps, or folds, and is also called the *tricuspid valve;* the left one has two cusps and is also called the *bicuspid,* or *mitral, valve.*

Atrium [ā′trē-um].

A chamber, such as the atrium of the heart, that receives blood from the sinus venosus or veins.

Auditory tube [aw′di-tōr-ē] (L. *audio,* p. p. *auditus* = to hear).

The tube that connects the tympanic (middle ear) cavity and the pharynx. It equalizes air pressure on each side of the tympanic membrane. Also called the *Eustachian tube.*

Auricle [aw′ri-kl] (L. *auricula* = little ear).

The external ear flap or pinna; also the ear-shaped lobe of a mammalian atrium.

Autonomic nervous system [aw-tō-nom′ik] (Gr. *autos* = self + *nomos* = rule, law).

An involuntary part of the nervous system (it "rules" itself.) It supplies motor fibers to glands and visceral organs.

Axillary [ak′sil-ār-ē] (L. *axilla* = armpit).

Pertaining to the armpit, e.g., the axillary artery.

Axis [ak′sis] (L. = axle, axis).

The second cervical vertebra of a mammal. Rotation of the head occurs between the atlas and axis.

Axon [ak′son] (Gr. = axle, axis).

The long, slender process of a neuron specialized to conduct nerve impulses a considerable distance.

Azygos vein [az′ī-gos] (Gr. *a* = without + *zygon* = yoke).

An unpaired vein in mammals. It drains most of the intercostal spaces on both sides of the body.

Basophil [bā′sō-fil] (Gr. *baso-* = alkaline + *phileo* = to love, to have an affinity for).

A leukocyte whose cytoplasmic granules take alkaline stains and appear blue; the rarest form of leukocytes.

Biceps [bī′seps] (L. *bi* = two + *ceps* = head).

A structure with two heads, such as the biceps brachii muscle, which is two-headed in human beings.

Bicornuate [bī-kōr′nū-āt] (L. *cornu* = horn).

A structure with two horns, such as a bicornuate uterus.

Bladder.

A membranous sac in which liquid may accumulate, e.g., the urinary bladder.

Blind spot.
See **Optic disk.**

Blood.
The liquid circulating in the blood vessels, consisting of a liquid plasma and cellular elements.

Bone.
The hard skeletal material of vertebrates consisting of an extracellular matrix of collagen fibers. It is mineralized with calcium phosphate crystals by bone forming cells.

Bowman's capsule. *(Sir William Bowman, British Anatomist, 1816–1832.)*
See **Glomerular capsule.**

Brachial [brā′kē-ăl] (L. *brachium* = upper arm).
Pertaining to the upper arm, e.g., brachial artery, coracobrachialis muscle.

Brachial plexus.
The network of nerves that supplies the shoulder and arm.

Brain.
The enlarged, cranial portion of the central nervous system. It is the major integrative center of the nervous system.

Braincase.
The cartilages and bones that encase the brain; also called the *cranium.*

Branchiomeric [brang′kē-o-mēr′ik] (Gr. *branchia* = gill + *meros* = part).
Pertaining to muscles or other structures associated with or derived from the visceral arches and gills.

Broad ligament.
Mesentery in female mammals. It anchors the reproductive tract to the dorsal body wall.

Bronchus [brong′kŭs] (Gr. *bronchos* = windpipe).
A branch of the trachea, which enters the lungs.

Buccal [bŭk′ăl] (L. *bucca* = cheek).
Pertaining to the mouth, e.g., buccal cavity.

Bulbourethral gland [bŭl′bō-yū-rē-thrăl] (L. *bulbus* = a bulbous root + Gr. *ourethra* = urethra).
A gland in male mammals near the base of the penis. It contributes to the seminal fluid. Also called *Cowper's gland.*

Caecum [sē′kŭm] (L. *caecus* = blind).
A pouch at the beginning of the large intestine. It is very long in many herbivores, including the rat, and houses bacteria and protozoa that digest cellulose.

Calcaneus [kal-kā′nē-ŭs] (L. = heel).
The large proximal tarsal bone that forms the "heel bone" in mammals.

Calyx, pl. **calyces** [kā′liks, kal′i-sēz] (Gr. *kaylx* = cup).
A cuplike compartment, such as one of the renal calyces, or subdivisions of the renal pelvis within the kidney.

Canaliculi [kan-ă-lik′yū-lī] (L. = small canals).
Minute canals in the bone matrix containing the slender processes of bone-forming cells.

Canine tooth [kā′nīn] (L. *canis* = dog).
The large pointed tooth in mammals. It is located caudal to the incisor teeth. In humans, the crown of the canine resembles an incisor, but its root is much larger. It is absent in the rat and enlarged in the pig.

Capillary [kap′i-lār-ē] (L. *capillus* = hair).
One of the minute, thin-walled blood vessels connecting small arteries to small veins. Exchanges between the blood and interstitial fluid occur across capillary walls.

Cardiac [kar′dē-ak] (Gr. *kardia* = heart).
Pertaining to the heart or its vicinity.

Carotid artery [ka-rot′id] (Gr. *karotides* = large neck artery, from *karoo* = to put to sleep, because compressing the artery leads to unconsciousness).
One of the arteries supplying blood to the head.

Carotid body.
A small enlargement at the junction of the external and internal carotid arteries. It is a chemoreceptor monitoring the levels of oxygen and carbon dioxide in the blood.

Carpal [kar′păl] (Gr. *karpos* = wrist).
One of the small bones forming the wrist.

Cartilage [kar′ti-lij] (L. *cartilago* = gristle).
A firm but elastic skeletal tissue whose matrix contains proteoglycan molecules that bind with water. Cartilage occurs in all embryos and in the skeleton of adult terrestrial vertebrates where firmness, smoothness, and flexibility are needed.

Caudal [kaw′dăl] (L. *cauda* = tail).
Pertaining to the tail or to a direction toward the tail in a quadruped.

Cecum.
See **Caecum.**

Cell body.
The part of a nerve cell, or neuron, that contains the nucleus. It does not include the axon and dendrites.

Central canal.
The cavity in the center of the spinal cord. It is continuous with the fourth ventricle in the brain and contains cerebrospinal fluid.

Central nervous system.
The part of the nervous system consisting of the brain and spinal cord.

Centrum.
See **Vertebral body.**

Cephalic [se-fal′ik] (Gr. *kephale* = head).
  Pertaining to the head, e.g., brachiocephalic artery.

Cerebellum [ser-ĕ-bel′um] (L. = small brain).
  The dorsal part of the metencephalon. It is an important center for muscular coordination and equilibrium.

Cerebral aqueduct [se-rē′brăl] (L. *cerebrum* = brain + *aqua* = water).
  The narrow passage within the brain connecting the third and fourth ventricles; also known as the *aqueduct of Sylvius.*

Cerebral hemisphere.
  One of two hemispheres that form most of the cerebrum. They are the major integration centers in the brain of mammals.

Cerebral peduncle [pe-dung′kl] (L. *pedunculus* = little foot).
  One of a pair of neuronal tracts, which can be seen on the ventral surface of the mammalian brain. It carries impulses from the cerebrum to motor centers in the brain stem and spinal cord.

Cerebrospinal fluid.
  A lymphlike fluid that circulates within and around the central nervous system and helps to protect and nourish it.

Cerebrum [se-rē′brŭm].
  The major part of the telencephalon. It includes the cerebral hemispheres and a group of deep nuclei.

Cervical [ser′vĭ-kal] (L. *cervix*, gen. *cervicis* = neck).
  Pertaining to the neck, e.g., cervical vertebrae.

Cervix [ser′viks].
  The neck of an organ, e.g., the cervix of the uterus.

Choana [kō′an-ă] (Gr. *choane* = funnel).
  One of the paired openings between the nasal cavity and pharynx; an internal nostril.

Choledochal duct.
  See **Common bile duct.**

Chorda tympani [kōr′dă tim-pan′ē] (L. = cord + Gr. *tympanon* = drum).
  A branch of the facial nerve crossing the malleus and traversing the tympanic cavity on its way to innervate certain taste buds on the tongue and salivary glands.

Chorion [ko′rē-on] (Gr. = skinlike membrane enclosing the fetus).
  The outermost extraembryonic membrane of reptiles, birds, and mammals.

Choroid [kō′royd] (Gr. *chorioeides* = resembling a chorion).
  The highly vascularized middle tunic of the eyeball lying between the retina and fibrous tunic.

Choroid plexus [plek′sŭs] (L. *plexus* = network).
  Vascular tufts that project from a thin layer of brain tissue and protrude into a ventricle of the brain. They secrete the cerebrospinal fluid.

Chromatophore [krō-mat′ō-fōr] (Gr. *chromo* = color + *phoros* = bearing, from *pherein* = to bear).
  A cell in nonmammalian vertebrates that contains pigment granules.

Ciliary body [sil′ē-ar-ē] (L. *cilium* = eyelash).
  A part of the vascular tunic of the eyeball attached to the lens. Its muscle fibers regulate accommodation, and its secretory cells produce the aqueous humor.

Clavicle [klav′i-kl] (L. *clavicula* = small key).
  The collar bone. It extends between the scapula and sternum in species in which it is well developed.

Cleido- [klī-do′] (Gr. *kleis*, gen. *kleidos* = key, clavicle).
  A root referring to the clavicle; used in combination with other terms, e.g., cleidobrachialis muscle.

Clitoris [klit′ō-ris] (Gr. *kleitoris* = hill).
  The small erectile organ of female mammals. It corresponds to much of the penis in a male mammal.

Cloaca [klō-ā′kă] (L. = sewer).
  A chamber in nonmammalian vertebrates that receives the termination of the digestive, urinary, and genital tracts.

Coagulating gland.
  A gland closely associated with the vesicular gland in male rats. It contributes to the seminal fluid.

Coccyx [kok′siks] (Gr. *kokkyx* = cuckoo).
  The several fused caudal vertebrae of human beings. As a whole, the coccyx resembles the shape of the bill of a cuckoo. It does not reach the body surface but serves as the attachment site for some muscles.

Cochlea [kok′lĕ-a] (L. = snail shell).
  Spiral part of the inner ear of mammals containing the auditory receptors and associated structures.

Coeliac artery [sē′lĕ-ak] (Gr. *koilia* = belly).
  A branch of the aorta that supplies the cranial abdominal viscera, including the spleen, stomach, and liver.

Coelom [sē′lom] (Gr. *koiloma* = a hollow).
  A body cavity that is completely lined by serosa, an epithelium of mesodermal origin.

Collagen [kol′lă-jen] (Gr. *kolla* = glue + *genos* = descent).
  Minute protein fibers that form most of the extracellular material in connective tissues.

Colliculus [ko-lik′yū-lŭs] (L. = little hill).
  One of four small elevations on the dorsal surface of the mammalian mesencephalon, which are centers for certain optic (superior colliculi) and auditory (inferior colliculi) reflexes.

Colon [kō′lon] (Gr. *kolon* = large intestine).
Most of the large intestine. Depending on the species, it extends from the small intestine or caecum to the cloaca or anus.

Columella [kol-yū-mel′a] (L. = small column).
A term frequently used for the rod-shaped stapes of nonmammalian terrestrial vertebrates. See also **Stapes.**

Commissure [kom′i-syūr] (L. *commissura* = a joining together).
A neuronal tract that interconnects structures on the left and right sides of the central nervous system.

Common bile duct.
The principal duct carrying bile to the intestine. It is formed by the confluence of hepatic ducts from the liver and, when present, the cystic duct from the gallbladder.

Concha [kon′kă] (Gr. *konkhe* = seashell).
One of several folds within the mammalian nasal cavity, which increase surface area; also called a turbinate bone.

Condyle [kon′dīl] (Gr. *kondylos* = knuckle).
A rounded articular surface, such as an occipital condyle.

Condyloid process.
The process of a mammalian mandible. It bears the facet for the jaw joint.

Conjunctiva [kon-jūnk-tī′va] (L. *conjunctus* = joined together).
The epithelial layer that covers the surface of, and fuses with, the cornea. It continues over the inner surface of the eyelids.

Connective tissue.
A widespread body tissue characterized by an extensive extracellular matrix containing fibers and cells. It includes collagenous and elastic connective tissue, fat, cartilage, and bone.

Conus arteriosus [kō′năs ar-ter-ē-ō′sas] (L. = cone).
A chamber of the heart in fishes and amphibians into which the single ventricle discharges blood. It contributes to the bases of the pulmonary trunk and aorta in mammals.

Coracoid [kōr′ă-koyd] (Gr. *korax*, gen. *korakos* = crow + *eidos* = appearance).
The bone forming the caudoventral part of the pectoral girdle in nonmammalian terrestrial vertebrates. It is reduced to a process (coracoid process) resembling a crow's beak in therian mammals.

Cornea [kōr′nē-ă] (L. *corneus* = horny).
The transparent part of the eyeball through which light passes. It is part of the fibrous tunic.

Coronary ligament [kōr′o-nār-ē] (L. *corona* = crown).
Peritoneum that bridges the gap between the liver and the diaphragm in mammals.

Coronary vessels.
Blood vessels that supply the heart musculature. During part of their course, they encircle the heart between the atria and ventricles.

Coronoid process.
The uppermost process of the mammalian mandible to which certain jaw muscles attach.

Corpus [kōr′pŭs] (L. = body).
The main body or part of an organ.

Corpus callosum [ka-lō′sum] (L. *callosus* = hard).
The large commissure interconnecting the two cerebral hemispheres in mammals.

Corpus cavernosum penis [kav-er-nō′sum pē′nis] (L. *caverna* = hollow space + *penis* = tail, penis).
One of a pair of columns of erectile tissue forming much of the penis.

Corpus luteum [lu-te′um] (L. *luteus* = yellow).
A yellowish endocrine gland within the ovary that develops from the ovulated follicle. The principal hormone it produces is progesterone.

Corpus spongiosum penis [spŭn-je′ō-sum] (Gr. *spongia* = sponge).
A column of erectile tissue that surrounds the penile portion of the urethra and forms the glans penis at the tip of the penis.

Cortex [kōr′teks] (L. = bark).
A layer of distinctive tissue on the surface of many organs, e.g., the cerebral cortex.

Costal [kos′tăl] (L. *costa* = rib).
Pertaining to the ribs, e.g., the costal cartilages.

Cowper's gland (*William Cowper*, British anatomist, 1666–1709).
See **Bulbourethral gland.**

Cranial [krā′nē-ăl] (Gr. *kranion* = skull).
Pertaining to the cranium; also a direction toward the head.

Cranium.
The skull, especially the part encasing the brain.

Cremasteric pouch [krē-mas-ter′ik] (Gr. *kremaster* = suspender).
Layers of the body wall that suspend the testis within the scrotum.

Crus, pl. **crura** [krūs, krū′ră] (L. = lower leg).
The lower leg or shin.

Cucullaris muscle [kyū′kū-lar-is] (L. *cucullus* = cap, hood).
A branchiomeric muscle in fishes and amphibians covering the craniodorsal part of the shoulder. It gives rise to the mammalian trapezius and sternocleidomastoid groups of muscles.

Cutaneous [kyū-tā′nē-ŭs] (L. *cutis* = skin).
Pertaining to the skin.

Cystic duct [sis′tik] (Gr. *kystis* = bladder).
The duct of the gallbladder. It joins the hepatic ducts to form the common bile duct.

Decussation [dē-kŭ-sā′shun] (L. *decusso*, p. p. *decussatus* = to divide crosswise in an X).

The crossing of neuronal tracts in the midline of the central nervous system.

Deferent duct.

See **Ductus deferens.**

Deltoid muscle [del′tōyd] (Gr. *deltoeides* = shaped like the letter delta, Δ).

A muscle crossing the lateral surface of the shoulder. It is shaped like the Greek letter *delta* in human beings.

Dendrite [den′drīte] (Gr. *dendron* = tree).

A highly branched and usually short process of a neuron that receives nerve impulses.

Dermis [der′mis] (Gr. *derma* = skin, leather).

The deeper layer of the skin. It is composed of dense connective tissue and develops from embryonic mesoderm.

Diaphragm [dī′ă-fram] (Gr. *dia* = through, across + *phragma* = partition, wall).

A mostly muscular partition between the thoracic and abdominal cavities in a mammal.

Diencephalon [dī-en-sef′ă-lon] (Gr. *enkephalos* = brain).

The brain region between the telencephalon and mesencephalon. It includes the epithalamus, thalamus, and hypothalamus.

Digit [dij′it] (L. *digitus* = finger).

A finger or toe.

Distal [dis′tăl] (L. *distalis* = situated away from the center).

The end of a structure most distant from its origin.

Dorsal [dōr′săl] (L. *dorsalis*, from *dorsum* = back).

A direction toward the surface of the back of a quadruped.

Duct [dŭkt] (L. *duco*, p. p. *ductus* = to lead).

A small, tubular passage carrying products away from an organ.

Ductus arteriosus [dŭk′tŭs ar-tēr′ē-ō-sŭs].

A connection in the fetal mammal between the pulmonary trunk and aorta, that permits much blood to bypass the embryonic lungs. It atrophies after birth and forms the ligamentum arteriosum.

Ductus deferens [def′er-enz] (L. *defero*, pres. p. *deferens* = to carry away).

The sperm duct of mammals and other amniotes; also called *vas deferens.*

Ductus venosus [vē-nō′sŭs].

A blood vessel in the liver of a fetal mammal. It permits much of the blood returning from the placenta in the umbilical veins to bypass the hepatic sinusoids and to enter the caudal vena cava directly.

Duodenum [dū-ō-dē′nŭm] (L. *intestinum duodenum digitorum, duodeni* = twelve fingers each).

The first part of the small intestine. In human beings, it is about 12 finger-breadths long.

Dura mater [dū-ră mā′ter] (L. = *durus* = hard + *mater* = mother).

The tough outer meninx surrounding the mammalian central nervous system.

Efferent ductules [ef′er-ent] (L. *ex* = out, away from + *fero*, pres. p. *ferens* = to carry).

Minute ducts in amphibians that carry sperm cells from the testis to the cranial tubules of the kidney; in amniotes, modified kidney tubules that lie in the head of the epididymis and transport sperm cells. These ducts are also called *vasa efferentia.*

Embryo [em′brē-ō] (Gr. *embryon* = embryo, from *en* = in + *bryo* = to swell).

An early stage in development of an individual. It is dependent for energy and nutrients from stored material within its egg or from the mother; i.e., embryos are not able to live on their own.

Endocrine gland [en′dō-krin] (Gr. *endon* = within + *krino* = to separate).

A ductless gland that discharges its secretion (a hormone) into the blood.

Eosinophil [ē-ō-sin′ō-fil] (Gr. *eos* = dawn + *philos* = fond).

A leukocyte whose cytoplasmic granules stain with eosin (an acid dye) and appear reddish.

Epaxial [ep-ak′sē-ăl] (Gr. *epi* = upon, above + *axon* = axle, axis).

Pertaining to those muscles and other organs that lie above or beside the dorsal half of the vertebral column.

Ependymal epithelium [ep-en′di-măl] (Gr. *ependyma* = garment).

The epithelial layer that lines the central canal of the spinal cord and the cavities in the brain.

Epidermis [ep-i-derm′is] (Gr. *epi* = upon + *derma* = skin).

The superficial layer of the skin. It is composed of a stratified, squamous epithelium whose outer layers are keratinized in terrestrial vertebrates. It is derived from the embryonic ectoderm.

Epididymis [ep-i-did′i-mis] (Gr. *didymoi* = testis).

A band-shaped group of tubules and a coiled duct, lying on the testis of mammals and in which sperm cells are stored. It evolved from a part of the kidney and kidney duct of ancestral amphibians and fishes.

Epiglottis [ep-i-glot′is] (Gr. *glottis* = entrance to the windpipe).

The flap of fibrocartilage that helps deflect food around the glottis of mammals during swallowing.

Epiphysis [e-pif′i-sis] (Gr. *physis* = growth).

(1) The end of a long bone in a young mammal.
(2) A thin outgrowth of the epithalamus in some fishes and amphibians, the distal end of which is sensitive to changes in light conditions. It corresponds to the mammalian pineal gland.

Glossary DEC–EPI

Epithalamus [ep′i-thal′ă-mŭs] (Gr. *thalamos* = inner chamber).
The roof of the diencephalon lying above the thalamus. Part of it is an olfactory center.

Epithelium [ep-i-thē′lē-um] (Gr. *thele* = delicate skin).
A tissue that is composed of tightly packed cells and covers body surfaces and lines cavities, including those of blood vessels and ducts.

Epitrichium [ep-i-trik′ē-ŭm] (Gr. *trichion* = small hair).
A layer of epithelium that lies upon the developing hairs in a mammal fetus.

Erectile tissue.
A tissue in an organ containing cavernous vascular spaces that may fill with blood and swell.

Esophagus [ē-sof′ă-gŭs] (Gr. *oisophagos* = gullet).
The part of the digestive tract between the pharynx and the stomach.

Eustachian tube (*Bartolommeo Eustachio*, Italian anatomist, 1520–1574).
See **Auditory tube.**

Exocrine gland [ek′sō-krin] (Gr. *ex* = out + *krino* = to separate).
A gland whose secretion is discharged through a duct onto a surface or into a cavity.

Extension [eks-ten′shŭn] (L. *extendo* = to stretch out).
A movement that carries a distal limb segment away from the next proximal segment, retracts a limb at the shoulder or hip, or moves the head or a part of the trunk toward the middorsal line.

External acoustic meatus [ă-kūs′tik mē-ā′tŭs] (Gr. *akoustikos* = pertaining to hearing + L. *meatus* = passage).
The external ear canal of amniotes extending from the body surface to the tympanic membrane.

External nostril.
See **Naris.**

Extrinsic [eks-trin′sik] (L. *extrinsecus* = from without).
A structure that originates from another structure or organ (e.g., extrinsic ocular muscles).

Extrinsic ocular muscles [ok′yū-lăr].
The group of small, strap-shaped muscles that extend from the wall of the orbit to the eyeball.

Facial [fā′shăl] (L. *facies* = face).
Pertaining to the face.

Facial nerve.
The seventh cranial nerve; innervates facial and other muscles associated with the second visceral arch and its derivatives, some salivary glands, and taste receptors on the front of the tongue.

Fallopian tube (*Gabriele Fallopio*, Italian anatomist, 1523–1562).
See **Uterine tube.**

Falx cerebri [falks se-rē′bri] (L. sickle + *cerebrum* = brain)
A sickle-shaped fold of dura mater in the longitudinal fissure between the two cerebral hemispheres.

120

Fascia [fash′ē-ă] (L. = band).
Sheets of connective tissue that lie beneath the skin (e.g., superficial fascia) or ensheathe groups of muscles (e.g., deep fascia) or organs.

Femur [fē′mŭr] (L. = thigh).
The thigh or bone within the thigh.

Fenestra [fe-nes′tră] (L. = window).
A windowlike opening in an organ.

Fenestra cochleae [kok′lĕ-ē] (L. *cochlea* = snail shell).
A small opening on the medial wall of the tympanic cavity of a mammal from which pressure waves, which have traveled through the inner ear, are released. It is often called the round window.

Fenestra vestibuli [ves-ti′būl-ē] (L. *vestibulum* = antechamber).
A small opening on the medial wall of the tympanic cavity into which the inner end of the stapes fits and propagates pressure waves into the inner ear. It is also called the *oval window.*

Fetus [fē′tŭs] (L. = offspring).
The unborn young of a mammal after it has nearly assumed the appearance it will have at birth (after the eighth week of gestation in human beings).

Fibroblast [fi′brō-blast] (L. *fibra* = fiber + Gr. *blastos* = bud).
An elongated and branching connective tissue cell that produces the intercellular matrix, including collagen fibers.

Fibrous tunic [tū′nik] (L. *tunica* = coat).
The tough connective tissue forming the outer wall of the eyeball; divided into the cornea and sclera.

Fibula [fib′yū-la] (L. = buckle, pin).
The slender bone on the lateral surface of the lower leg.

Fissure [fish′ūr] (L. *fissura* = cleft).
A deep groove or cleft in an organ, such as the brain or skull.

Flexion [flek′shŭn] (L. *flexio* = the bending).
A movement that brings a distal limb segment toward the next proximal one, advances a limb at the shoulder or hip, or bends the head or a part of the trunk toward the midventral line of the body.

Folia [fō′lē-ă] (L. = leaves).
Leaflike folds in an organ, such as those in the cerebellar hemispheres.

Foramen, pl. foramina [fō-rā′men, fō-ram′i-nă] (L. = opening).
An opening in an organ.

Foramen magnum [mag′nŭm] (L. *magnus* = large).
The opening in the base of the skull through which the spinal cord passes.

Foramen of Monro (*Alexander Monro, Sr.*, Scottish anatomist, 1697–1767).
See **Interventricular foramen.**

Foramen ovale [ō-vāl′ē] (L. *ovalis* = oval).

A valved opening in the interatrial septum of a fetal mammal. It permits much of the blood (chiefly blood from the placenta and caudal part of the body) to pass from the right to the left atrium, thus, bypassing the lungs. It closes at birth and becomes the adult fossa ovalis.

Forebrain.

See **Prosencephalon.**

Fossa [fos′ă] (L. = ditch).

A shallow groove or depression in an organ.

Fossa ovalis [ō-vah′lis] (L. *ovalis* = oval).

An oval depression in the medial wall of the right atrium of an adult mammal. It is a vestige of the fetal foramen ovale.

Frontal [frŭn′tăl] (L. *frons*, gen. *frontis* = forehead).

(1) Pertaining to the forehead. (2) A plane of the body that passes through the frontal suture between the frontal and parietal bones; i.e., a median longitudinal plane passing from left to right.

Gallbladder [gawl′blad-er] (Old English *galla* = bile).

A small sac attached to the liver in most vertebrates. It stores and concentrates bile. The gallbladder is absent in rats.

Gametes [gam′ētz] (Gr. *gamete* = wife, *gametes* = husband).

The haploid germ cells, eggs and sperm.

Ganglion [gang′glē-on] (Gr. = small tumor, swelling).

A group of neuronal cell bodies that is a part of the peripheral nervous system in vertebrates.

Gastric [gas′trik] (Gr. *gaster* = stomach).

Pertaining to the stomach.

Gemelli muscles [jē-mel′lĭ] (L. = small twins).

Two small muscles situated deeply on the lateral surface of the hip in many mammals. They are fused with the obturatorius muscle in the pig.

Genio- [jē-ni′ō] (Gr. *geneion* = chin).

A combining term pertaining to the chin, as in the genioglossus muscle.

Girdles.

The skeletal elements in the body wall of vertebrates that support the appendages.

Gland (L. *glans* = acorn).

A group of secretory cells. Exocrine glands discharge their secretion by a duct onto the body surface or into a cavity. Ductless endocrine glands discharge their secretion into the blood.

Glans clitoridis [glanz kli-tō′ri-dis] (Gr. *kleitoris* = hill).

The small mass of erectile tissue at the distal end of the clitoris in a female mammal.

Glans penis [pē′nis] (L. *penis* = tail, penis).

The bulbous, distal end of the penis in a mammal; part of the corpus spongiosum penis.

Glenoid fossa [glen′oyd] (Gr. *glene* = socket + *eidos* = resembling).

The socket that is located on the pectoral girdle of terrestrial vertebrates and receives the head of the humerus.

Glomerular capsule [glō-măr′yū-lăr] (L. *glomerulus* = little ball of yarn).

The thin-walled, expanded proximal end of a renal tubule. It surrounds a tuft of capillaries, the glomerulus. It is also called *Bowman's capsule.*

Glomerulus [glō-măr′yū-lŭs] (L. = little ball of yarn).

A dense network of capillaries that is surrounded by the glomerular capsule at the proximal end of a kidney tubule.

Glossopharyngeal nerve [glos′ō-fă-rin-jē-ăl] (Gr. *glossa* = tongue + *pharynx* = throat).

The ninth cranial nerve. It carries motor fibers to pharyngeal muscles derived from the third visceral arch, parasympathetic fibers to certain salivary glands, and returns sensory fibers from taste buds on part of the tongue and from the part of the pharynx near the base of the tongue.

Glottis [glot′is] (Gr. = opening of the windpipe).

A slitlike opening in the center of the larynx, including the space between the vocal cords.

Gluteal [glū′tē-ăl] (Gr. *gloutos* = buttock).

Pertaining to the buttocks, e.g., gluteal muscles.

Goblet cells.

Goblet-shaped, mucus-producing cells associated with the epithelial lining of the stomach, intestine, and upper respiratory tract.

Gonads [gō′nadz] (Gr. *gone* = seed).

A collective term for the testes and ovaries.

Graafian follicle [gră′fē-ăn] (*Reijnier de Graaf*, Dutch anatomist, 1641-1673).

See **Tertiary follicle.**

Gray matter.

Tissue in the central nervous system consisting primarily of cell bodies of neurons and unmyelinated nerve fibers.

Gubernaculum [gū′ber-nak′yū-lŭm] (L. = small rudder).

A cord of connective tissue that lies at the caudal end of the testis and guides the descent of the testis into the scrotum.

Gyrus [jī′rŭs] (Gr. *gyros* = circle).

One of the folds on the surface of a cerebral hemisphere.

Hair.

A filamentous epidermal structure in the skin of mammals. It consists primarily of keratinized cells. En masse, it helps to thermally insulate the body.

Hallux [hal′ŭks] (Gr. = big toe).

The first or most medial digit of the foot.

Harderian gland (*Johann Harder,* Swiss anatomist, 1656-1711).
A tear gland present in some mammals. It is located rostral and ventral to the eyeball. Also called *accessory lacrimal gland.*

Hard palate.
The part of the secondary palate that is supported by a horizontal shelf of bone in the mammalian skull and separates the oral from the nasal cavities.

Haversian system (*Clopton Havers,* British anatomist, 1650-1702).
See **Osteon.**

Heart.
The hollow muscular organ that pumps blood through the body.

Hemoglobin [hē-mō-glō′bin] (Gr. *haima* = blood + L. *globus* = globe).
Iron-containing, globular molecules that fill the red blood cells and bond reversibly with oxygen and some carbon dioxide.

Hepatic [he-pat′ik] (Gr. *hepar* = liver).
Pertaining to the liver.

Hepatic duct.
One of the ducts that drain the liver. Hepatic ducts join a cystic duct from the gallbladder to form the common bile duct in animals that possess a gallbladder.

Hepatic portal system.
A system of veins that drain the capillaries of the abdominal digestive organs and spleen and that lead to the sinusoids within the liver.

Hepatic vein.
One of the veins that drains the hepatic sinusoids and, in mammals, leads to the caudal vena cava.

Hindbrain.
See **Rhombencephalon.**

Hyaloid artery [hi′ă-loyd] (Gr. *hyolos* = glass + *eidos* = appearance).
The embryonic artery that passes through the vitreous body of the eyeball to supply the developing lens. It disappears before birth.

Hyobranchial apparatus [hī′ō-brang′kē-ăl] (Gr. *hyo* = a combining form referring to structures associated with the second visceral, or hyoid, arch + *branchia* = gill).
The complex of cartilages and bones that is derived from the hyoid and other visceral arches and which supports the tongue and the floor of the mouth and pharynx in nonmammalian terrestrial vertebrates. It is also called the hyoid apparatus in mammals.

Hyoid [hī′oyd] (Gr. *hyoeides* = resembling the letter *upsilon* = U-shaped).
(1) Pertaining to structures associated with the second visceral, or hyoid, arch. (2) The

mammalian bone that is embedded in and supports the base of the tongue.

Hyoid apparatus.
See **Hyobranchial apparatus.**

Hypaxial [hī-pak′sē-ăl] (Gr. *hypo* = under + *axon* = axle or axis).
Pertaining to those muscles and other structures that lie in the body wall ventral to the dorsal half of the vertebral axis.

Hypobranchial [hī-pō-brang′kē-ăl] (Gr. *branchia* = gill).
Pertaining to muscles and other structures derived from structures that are located ventral to the gills in fishes.

Hypoglossal nerve [hī-pō-glos′ăl] (Gr. *glossa* = tongue).
The twelfth cranial nerve of amniotes. It carries motor fibers to the muscles of the tongue.

Hypophysis [hī-pof′i-sis] (Gr. *physis* = growth).
An endocrine gland that is attached to the ventral surface of the hypothalamus. It secretes many hormones that regulate a variety of physiological processes and other endocrine glands. Also called the pituitary gland.

Hypothalamus [hī-pō-thal′ă-mŭs] (Gr. *thalamus* = inner chamber).
The ventral part of the diencephalon. It lies ventral to the thalamus and is an important center for the integration of visceral functions.

Ileum [il′ē-ŭm] (L. = small intestine).
The caudal half of the small intestine of mammals. In human beings it is the part of the small intestine that lies against the ilium.

Iliac [il′ē-ak] (L. *ilium* = groin, flank).
Pertaining to a structure near the groin or flank, e.g., iliac artery.

Ilium [il′ē-ŭm] (L. = groin or flank).
The bone forming the dorsal part of the pelvic girdle of terrestrial vertebrates. It articulates with the sacrum.

Incisor tooth [in-ŝi′zŏr] (L. = the cutter, from *incido,* p. p. *incisum* = to cut into).
One of the front teeth of mammals lying rostral to the canine tooth. It is used for cutting and cropping and is very large in rodents.

Incus [ing′kŭs] (L. = anvil).
The anvil-shaped, middle auditory ossicle of mammals.

Inferior [in-fē′rē-or] (L. = lower).
A direction toward the feet of a human being.

Infundibulum [in-fŭn-dib′yū-lŭm] (L. = little funnel).
A funnel-shaped structure, such as the entrance into the oviduct or uterine tube.

Inguinal [ing′gwi-năl] (L. *inguen,* gen. *inguinis* = groin).
Pertaining to structures in or near the groin.

Inguinal canal.
A passage through the muscular and fascial layers of the body wall, leading from the abdominal cavity to the vaginal cavity of the scrotum. Ducts, blood vessels, and nerves to and from the testis travel through this passage.

Inner ear.
The portion of the ear that lies within the otic capsule of the skull and contains the receptive cells for equilibrium and hearing.

Insertion of a muscle.
The point of attachment of a muscle that moves the greater distance when the muscle contracts. It is usually the distal end of a limb muscle.

Integument [in-teg'yū-ment] (L. *integumentum* = covering).
The body covering consisting of the epidermis and the dermis. It often includes many accessory structures: glands, hair, feathers, scales. Also called skin.

Internal nostril.
See **Choana.**

Interneuron [in'ter-nū-ron] (L. *inter* = between + Gr. *neuron* = nerve, sinew).
A neuron that lies within the central nervous system and connects sensory and motor neurons or other interneurons. It is responsible for most of the integrative activity of the nervous system.

Interstitial cells [in-ter-stish'al] (L. *interstitium* = a small space between).
Groups of endocrine cells that lie between the seminiferous tubules of the testis and secrete the male sex hormone, testosterone.

Interstitial fluid.
A lymphlike fluid that lies in the minute spaces between the cells of the body.

Interthalamic adhesion [in-ter-tha-lam'ik] (Gr. *thalamos* = inner chamber).
The part of the mammalian thalamus that crosses the midline within the third ventricle. Also called the massa intermedia.

Interventricular foramen [in-ter-ven-trik'yū-lar fō-ra'men] (L. *ventriculus* = belly).
A foramen leading from one of the lateral ventricles of the brain to the third ventricle. Also called the *foramen of Monro.*

Intestine [in-test'tin] (L. *intestinus* = intestine).
The portion of the digestive tract between the stomach and anus or cloaca. It is the site of most digestion and absorption.

Intrinsic [in-trin'sik] (L. *intrinsecus* = on the inside).
A structure that is an inherent part of an organ, e.g., the ciliary muscles of the eyeball.

Iris [ī'ris] (Gr. = rainbow).
The pigmented part of the vascular tunic of the eyeball. It surrounds the pupil.

Ischiadic, ischiatic [is-ke-ad' (at) -ik] (Gr. *ischion* = hip).
Pertaining to a structure near the hip, e.g., the ischiadic nerve.

Ischium [is'kē-ŭm].
The bone forming the caudoventral part of the pelvic girdle.

Islets of Langerhans (*Paul Langerhans*, German physician, 1847–1888).
See **Pancreatic islets.**

Jacobson's organ (*Ludwig L. Jacobson*, Danish surgeon and anatomist, 1783–1843).
See **Vomeronasal organ.**

Jejunum [jĕ-jū'nŭm] (L. *jejunus* = empty).
Approximately the first half of the mammalian postduodenal small intestine. It is usually found to be empty in autopsies.

Jugular vein [jug'yū-lar] (L. *jugulum* = throat).
One of the major veins in the neck of mammals. It drains the head.

Keratin [ker'ă-tin] (Gr. *keras* = horn).
A horny protein synthesized by the epidermal cells of terrestrial vertebrates.

Kidney [kid'nē] (Middle English *kindenei* = kidney).
The organ that removes waste products, especially nitrogenous wastes, from the blood and produces urine.

Kidney tubule.
See **Renal tubule.**

Kneecap.
See **Patella.**

Labia [lā'bē-ă] (L. = lips).
Liplike structures, e.g., the genital labia of a female mammal.

Lacrimal [lak'ri-măl] (L. *lacrima* = tear).
Pertaining to structures involved with the production and transport of tears (e.g., lacrimal gland) or to structures located near these structures (e.g., lacrimal bone).

Lacuna [lă-kū'nă] (L. = a small hollow or space).
A small cavity, such as one of many in the bone matrix, in which a bone cell (osteocyte) lodges.

Lamella [lă-mel'ă] (L. = small plate, layer).
A thin plate or layer of tissue, such as the layers of matrix in bone.

Laryngotracheal chamber.
A chamber in the respiratory tract of amphibians into which the glottis leads and from which the lungs emerge. It is comparable to the larynx and trachea of mammals.

Larynx [lar'ingks] (Gr. = larynx).
The part of the respiratory tract of amniotes between the pharynx and trachea. It contains the vocal cords in mammals.

Lateral [lat'er-ăl] (L. *latus*, gen. *lateris* = side).
Pertaining to the side of the body.

Lens [lenz] (L. = lentil).
A refractive transparent body near the front of the eyeball. It is responsible for accommodation or focusing by changing its shape (mammals) or by moving toward or away from the retina (frogs).

Leukocyte [lū'kō'sīt] (Gr. *leukos* = clear, white + *kytos* = cell).
Any of several different types of white blood cells.

Lienic [lī'ĕ-nik] (L. *lien* = spleen).
Pertaining to the spleen, e.g., lienogastric artery.

Ligament [lig'ă-ment] (L. *ligamentum* = band, bandage).
(1) A band of dense connective tissue extending between certain structures, usually bones. (2) A mesentery extending between certain visceral organs.

Ligamentum arteriosum [lig'ă-men-tum ar-tēr'ē-ō-sum].
A band of dense connective tissue extending from the pulmonary trunk to the aorta in adult mammals. It is a remnant of the embryonic ductus arteriosus.

Linea alba [lin'ē-ă al'ba] (L. = white line).
A whitish strand of connective tissue in the midventral abdominal wall to which abdominal muscles attach.

Limbic system [lim'bik] (L. *limbus* = border).
A brain region that encircles the diencephalon and leads to the hypothalamus. It is important in regulating many behaviors related to species survival, including feeding and reproduction.

Lingual [ling'gwăl] (L. *lingua* = tongue).
Pertaining to the tongue.

Liver [liv'er] (Old English *lifer* = liver).
The large organ in the cranial part of the abdominal cavity. It secretes bile and plays a vital role in many metabolic processes, including processing many substances brought to it by the hepatic portal vein system, disassembling senescent red blood cells, detoxifying substances, and synthesizing many plasma proteins.

Lumbar [lŭm'bar] (L. *lumbus* = loin).
Pertaining to structures in the back between the thorax and pelvis, such as the lumbar vertebrae.

Lung [lŭng] (Old English *lungen* = lung).
One of the paired organs of terrestrial vertebrates in which gases are exchanged between the air and blood. It develops as an outgrowth from the floor of the pharynx.

Lymph [limf] (L. *lympha* = clear water).
A clear liquid that is derived from the interstitial fluid and flows in the lymphatic vessels. It lacks the red blood cells and most of the plasma proteins found in blood.

Lymph heart.
A pulsating part of lymphatic vessels of some amphibians and reptiles. It helps return lymph to the veins.

Lymph node.
Nodules of lymphatic tissue along the course of the lymphatic vessels of mammals. It is the site for the multiplication of many lymphocytes, the phagocytosis of many foreign particles, and the initiation of certain immune response.

Lymphocyte [lim'fō-sīt] (Gr. *kytos* = cell).
A leukocyte with a large, round nucleus and very little cytoplasm. Lymphocytes develop in lymph nodes, spleen, thymus, and other lymphoid tissues, and participate in immune responses.

Macrophage [mak'rō-fāj] (Gr. *makros* = large + *phagein* = to eat).
Large cell in the tissues. It phagocytoses foreign particles and participates in immune responses.

Malleus [mal'ē-ŭs] (L. = hammer).
The outermost of three auditory ossicles of a mammal. It transmits sound waves across the tympanic cavity; it is shaped like a hammer.

Mammal [mam'ăl] (L. *mamma* = breast).
A member of the vertebrate class characterized by hair and mammary glands.

Mammary gland [mam'ă-rē].
The milk-secreting, cutaneous gland that characterizes female mammals.

Mammary papilla [pă-pil'ă] (L. = nipple, pimple).
The nipple, or teat, of a mammary gland.

Mandibular cartilage [man-dib'yū-lăr] (L. *mandibula* = lower jaw).
A cartilaginous core of the bony lower jaw of many fishes, amphibians, and reptiles. It represents the ventral half of the first visceral arch of ancestral fishes.

Mandibular gland.
A mammalian salivary gland usually located deep to the caudoventral angle of the lower jaw.

Mandibular ramus [rā'mŭs] (L. = a branch).
The vertical part of the mandible caudal to the teeth.

Massa intermedia.
See **Interthalamic adhesion.**

Masseter muscle [mă-sē'ter] (Gr. *masseter* = chewer).
A large mammalian muscle that helps to close the jaws. It extends from the zygomatic arch to the mandible.

Mastoid process [mas′toyd] (Gr. *mastos* = breast + *eidos* = appearance).

A large, rounded process on the base of the mammalian skull located caudal to the external acoustic meatus; it is a point of attachment for certain neck muscles.

Meatus [mē-ā′tŭs] (L. = a passage).

A passage, such as the external acoustic meatus, that leads from the head surface to the tympanic membrane.

Meconium [mē-kō′nē-um] (Gr. *mekonion* = poppy juice).

Bile-stained debris in the fetal digestive tract. It is discharged shortly after birth.

Medial [mē′dē-al] (L. *medialis* = middle).

A direction toward the middle of the body.

Median [mē′dē-an] (L. *medianus* = middle).

Lying in the midline of the body.

Mediastinum [me′dē-as-tī-nŭm] (L. = middle septum).

The space in the mammalian thorax between the two pleural cavities. It contains the aorta, esophagus, pericardial cavity, heart, and the venae cavae.

Medulla [me-dūl′ă] (L. = core, marrow).

The central part, or core, of certain organs, as opposed to their surface region, or cortex.

Medulla oblongata [ob-long-gah′tă].

The myelencephalon, or caudal region of the brain, which is continuous with the spinal cord.

Melanin [mel′ă-nin] (Gr. *melas* = black).

A black pigment often contained in cells known as melanophores.

Meninx, pl. **Meninges** [mē′ningks, mě-nin′jēz]
(Gr. = membrane).

A connective tissue membrane that surrounds the central nervous system.

Mesencephalon [mez-en-sef′ă-lon] (Gr. *mesos* = middle + *enkephalos* = brain).

The middle, or third, of five brain regions. It lies between the diencephalon and metencephalon and includes the colliculi (mammals) or optic lobes (frogs); also called the midbrain.

Mesentery [mez′en-ter-ē] (Gr. *enteron* = intestine).

(1) Any membrane-like double layer of serosa that extends from the body wall to visceral organs, or between visceral organs. (2) The particular membrane suspending the intestine.

Mesonephric duct.

See **Archinephric duct.**

Mesorchium [mez-ōr′kē-ŭm] (Gr. *orchis* = testis).

The mesentery suspending the testis.

Mesovarium [mez′ō-va′rē-um] (L. *ovarium* = ovary).

The mesentery suspending the ovary.

Metacarpal [met′ă-kar′păl] (Gr. *meta* = after + *karpos* = wrist).

One of the long bones in the palm of the hand located between the carpals and phalanges.

Metatarsal [met′ă-tar′săl] (Gr. *tarsos* = ankle).

One of the long bones of the foot located between the tarsals and phalanges.

Metencephalon [met′en-sef′ă-lon] (Gr. *enkephalos* = brain).

The fourth of the five brain regions. It lies between the mesencephalon and myelencephalon and includes the cerebellum and (in mammals) the pons.

Midbrain.

See **Mesencephalon.**

Middle ear.

The portion of the ear of terrestrial vertebrates that usually contains the tympanic cavity and one or three auditory ossicles. Airborne sound waves are transmitted from the body surface (usually from a tympanic membrane) across the middle ear cavity via the auditory ossicle (or ossicles) to the inner ear.

Middle ear cavity.

See **Tympanic cavity.**

Molar tooth [mō′lăr] (L. *molaris* = millstone).

One of the caudal teeth of mammals usually adapted for crushing or grinding food.

Monocyte [mon′ō-sīt] (Gr. *monos* = single + *kytos* = cell).

A leukocyte with a kidney-shaped nucleus. It is a precursor of a tissue macrophage.

Mouth [mowth] (Old English *muth* = mouth).

The cranial opening of the digestive tract. Also used for the oral cavity into which this opening leads.

Mucosa [myū-kō-să] (L. *mucosus* = slimy, mucous).

The lining of the digestive tract and many other hollow visceral organs. It consists of an epithelium, connective tissue, and sometimes a thin layer of smooth muscle.

Mucus [myū′kŭs] (L. = slime).

A slimy material secreted by some epithelial cells. It is rich in the glycoprotein mucin.

Muscle [mŭs-ěl] (L. *musculus* = muscle, a little mouse).

(1) A contractile tissue responsible for most of the movements of the body or its parts. (2) A discrete group of muscle fibers with a common origin and insertion.

Myelencephalon [mī′el-en-sef′ă-lon] (Gr. *myelos* = core, spinal cord + *enkephalos* = brain).

The most caudal of the five regions of the brain. It consists of the medulla oblongata.

Mylohyoid [mī′lō-hī′ōyd] (Gr. *myle* = a mill + *hyoeides* = U-shaped).

A nearly transverse sheet of muscle extending between the two mandibles and the hyoid bone.

*Glossary
MAS–MYL*

125

Myo- [mī′ō] (Gr. *mys*, gen. *myos* = muscle).
A prefix meaning muscle or musclelike.

Myoepithelium [mī′ō-ep-thē-lē-um].
Specialized epithelial cells containing contractile elements. Some surround the secretory cells of sweat glands and help to discharge sweat.

Myofibrils [mī-ō-fī′brilz] (L. *fibrilla* = a minute fiber).
Minute, longitudinal, contractile fibrils within a muscle fiber. They are barely visible with a light microscope and are composed of myofilaments.

Myofilaments [mī′ō-fil′ă-mentz].
Ultramicroscopic filaments of actin and myosin whose interactions are the basis of muscle contraction.

Naris, pl. **nares** [nā′ris, nā′res] (L. = nostril).
An opening from the outside into the nasal cavity; an external nostril.

Nasal [nā′zăl] (L. *nasus* = nose).
Pertaining to the nose, e.g., nasal bone, nasal cavity.

Nasal conchae [kon′kē] (L. = shells).
Folds within the nasal cavity of mammals. They increase the surface area.

Nasal meatus (L. = a passage).
An air passage in the mammalian nasal cavity between the nasal conchae, or between the conchae and nasal septum.

Nephron [nef′ron] (Gr. *nephros* = kidney).
The structural and functional unit of the kidney. It consists of a glomerulus and a renal tubule.

Nerve [nerv] (L. *nervus* = nerve, sinew).
A bundle of neuronal processes (axons) and their investing connective tissues. It extends between the central nervous system and peripheral organs, and is part of the peripheral nervous system.

Neural arch.
See **Vertebral arch.**

Neuron [nūr′on] (Gr. *neuron* = nerve, sinew).
A nerve cell. It is the structural and functional unit of the nervous system. It consists of dendrites, a cell body, and an axon.

Neuron tract.
A bundle of neuronal processes within the spinal cord or brain.

Neutrophile [nū′trō-fil] (L. *neutro* = neither + Gr. *philos* = affinity for).
A leukocyte with small cytoplasmic granules that stain only lightly with basic and acidic dyes. Neutrophiles are phagocytic and are the most abundant leukocytes.

Nictitating membrane [nik′ti-tāt-ing] (L. *nicto* = to beckon, wink).
A membrane in many terrestrial vertebrates that can slide across the surface of the eyeball. It is

reduced to a vestigial semilunar fold in human beings.

Nipple [nip′l] (Old English *neb* = beak, nose).
A papilla that bears the openings of the ducts from the mammary gland.

Nostril [nos′tril] (Old English *nosus* = nose + *thyrl* = hole).
The opening into the nasal cavity from the body surface (external nostril, naris) or from the nasal cavity into the pharynx (internal nostril, choana).

Obturator foramen [ob′tū-rā-tŏr] (L. *obturo*, p. p. *obturatus* = to close by stopping up).
A large foramen in the mammalian pelvic girdle. The obturator muscles arise from the periphery of the foramen and close it.

Occipital [ok-sip′i-tăl] (L. *occiput* = back of the head).
Pertaining to the back of the head or skull.

Occipital condyle [kon′dīl] (Gr. *kondylos* = condyle, knuckle).
One of a pair of enlargements on the occipital bone on each side of the foramen magnum of amphibians and mammals, or a single enlargement ventral to the foramen magnum in most other vertebrates. They articulate with the cranial articular surface of the atlas.

Ocular [ok′yū-lăr] (L. *oculus* = eye).
Pertaining to the eye, e.g., the extrinsic ocular muscles that move the eyeball.

Oculomotor nerve [ok′yū-lō-mō-tŏr].
The third cranial nerve. It carries motor nerve fibers to most of the extrinsic muscles of the eyeball and autonomic nerve fibers into the eyeball.

Olecranon [ō-lek′ră-non] (Gr. = the tip of the elbow).
The proximal end of the ulna. It extends behind the elbow joint in mammals.

Olfactory [ol-fak′tŏ-rē] (L. *olfacio*, p. p. *olfactus* = to smell).
Pertaining to the parts of the nose involved in smelling.

Olfactory nerve.
The first cranial nerve, which consists of neurons returning from the nose to the olfactory bulb of the brain.

Omentum [ō-men′tŭm] (L. = fatty membrane).
The peritoneal membrane, often containing a great deal of fat. It extends between the body wall and stomach (i.e., greater omentum) or between the stomach and the liver and duodenum (i.e., lesser omentum).

Omo- [ō′mō] (Gr. *omos* = shoulder).
A combining form pertaining to the shoulder, e.g., omotransversarius muscle.

Oocyte [ō′ō-sīt] (Gr. *oon* = egg + *kytos* = cell).
An early stage in the development of the egg. The first meiotic division of the primary oocyte produces the secondary oocyte and a polar body.

Oogonium [ō-ō-gō′nē-ŭm] (Gr. *gone* = generation).
> A very early stage in the development of an egg. It enlarges to become the primary oocyte.

Ootid [ō-ō-tid] (Gr. *ootidium* = a small egg).
> The nearly mature egg, or ovum, after the second meiotic division has been initiated. In mammals, fertilization initiates the completion of this division.

Optic [op′tik] (Gr. *optikos* = pertaining to the eye).
> Pertaining to the parts of the eye involved in vision.

Optic chiasma [kī-az′ma] (Gr. = cross, from the Greek letter *chi* = χ).
> The complete or partial decussation of the optic nerve fibers on the ventral surface of the diencephalon.

Optic disk.
> A disk-shaped area on the retina to which the optic nerve attaches. It lacks photoreceptors; also called the "blind spot."

Optic lobe.
> One of a pair of lobes on the dorsal surface of the mesencephalon in nonmammalian vertebrates. It is a major integrating center in these vertebrates.

Optic nerve.
> The second cranial nerve. It carries impulses from the retina to the brain.

Optic tract.
> The neuronal tract leading from the optic chiasma to the thalamus, or optic lobes, or both.

Oral cavity [ōr′ăl] (L. *os*, gen. *oris* = mouth).
> The mouth cavity; also called the buccal cavity.

Orbit [or′bit] (L. *orbis*, gen. *orbitis* = circle).
> A circular cavity on the side of the skull. It lodges the eyeball.

Origin of a muscle.
> The point of attachment of a muscle that tends to remain in a fixed position when the muscle contracts; the proximal end of a limb muscle.

Ossicle [os′i-kl] (L. *ossiculum* = small bone).
> Any small bone, such as one of the auditory ossicles.

Osteocyte [os′tē-ō-sīt] (Gr. *osteon* = bone + *kytos* = cell).
> A mature bone cell surrounded by the matrix it has produced.

Osteon [os′tē-on].
> A cylindrical, microscopic unit of bone consisting of concentric layers of bone matrix surrounding a central canal that contains blood and lymph vessels. Also called a *Haversian system.*

Ostium [os′tē-ŭm] (L. = entrance, mouth).
> The entrance to an organ, such as the ostium tubae of the uterine tube.

Otic capsule [ō′tik] (Gr. *otikos* = pertaining to the ear).
> The part of the skull surrounding the inner ear.

Otolith [ō′tō-lith] (Gr. *oto-* = ear + *lithos* = stone).
> Calcareous granules within sacs of the inner ear that stimulate sensory cells over which they move when an animal moves or changes position.

Oval window.
> See **Fenestra vestibuli.**

Ovarian follicle [ō-var′ē-an] (L. *ovarium* = ovary + *folliculus* = little bag).
> A group of epithelial and connective tissue cells in the ovary. It surrounds and nourishes the developing egg. It is also an endocrine gland whose primary hormone is estrogen.

Ovary [ō′vă-rē].
> One of a pair of female reproductive organs containing the ovarian follicles and eggs.

Oviduct [ō′vi-dŭkt] (L. *ovum* = egg + *ducere*, p. p. *ductus* = to lead).
> The passage in females of nonmammalian vertebrates. It transports eggs from the coelom to the cloaca.

Oviparous [ō-vip′ă-rŭs] (L. *pario* = to bear).
> To bear eggs. A pattern of reproduction found in frogs and many nonmammalian vertebrates that lay eggs. The embryos develop outside the body of the mother.

Ovisac [ō′vi-sak].
> An enlargement at the caudal end of the oviduct in which eggs accumulate before they are released.

Ovulation [ov′yū-lā-shŭn].
> The release of mature egg cells from the ovarian follicles and ovary into the coelomic cavity (in the frog), ovarian bursa (in the rat and pig), or infundibulum (in the human female and many other mammals).

Ovum [ō′vum].
> The mature egg cell.

Palate [pal′ăt] (L. *palatum* = palate).
> The roof of the mouth. See also **Secondary palate.**

Pampiniform plexus [pam-pin′i-fōrm] (L. *pampinus* = tendril + forma = shape).
> A network of veins in mammals entwining the testicular artery as it approaches the testis.

Pancreas [pan′krē-as] (Gr. *pan* = all + *kreas* = flesh).
> A large glandular outgrowth of the duodenum. It secretes many digestive enzymes. It also contains the endocrine pancreatic islets of Langerhans.

Pancreatic islets.
> Small clusters of endocrine cells in the pancreas. They produce hormones that regulate sugar metabolism; also called the *islets of Langerhans.*

Papilla [pă-pil′ă] (L. = nipple).
> A small conical protuberance.

Papilla amphibiorum.
> The part of the amphibian inner ear receptive to low-frequency sound waves.

Paradidymis [par′ă-did′i-mis] (Gr. *para* = beside + didymoi = testis).
> A small group of vestigial kidney tubules occurring in male mammals and located beside the epididymis.

Paraflocculus [par-ă-flok′yū-lŭs] (L. *flocculus* = a small tuft of wool).
> A small lobe on the lateral surface of the cerebellum. It is conspicuous in the rat.

Parasympathetic [par-ă-sim-pa-thet′ik] (Gr. *syn* = with + *pathos* = feeling).
> Pertaining to the parasympathetic part of the autonomic nervous system. In mammals, preganglionic parasympathetic neurons leave the central nervous system through certain cranial nerves and sacral spinal nerves. The parasympathetic system stimulates metabolic processes that absorb and store energy.

Parathyroid gland [par-ă-thī′royd] (Gr. *thyreos* = oblong shield + *eidos* = shape).
> One of several endocrine glands of a terrestrial vertebrate. It is located dorsal to, near, or entwined with the thyroid gland. Its hormone, called parathormone, helps to regulate calcium and phosphate metabolism.

Parietal [pă-rī′ĕ-tăl] (L. *paries*, gen. *parietis* = wall).
> Pertaining to the wall of some structure, e.g., parietal bone, parietal peritoneum.

Parotid gland [pă-rot′id] (Gr. *para* = beside + otikos = pertaining to the ear).
> A large mammalian salivary gland located ventral to the external ear.

Patella [pa-tel′ă] (L. = small plate).
> The mammalian kneecap, composed of bone. It is a sesamoid bone within the patellar tendon of the quadriceps muscle.

Pectoral [pek′tŏ-răl] (L. *pectoralis* = pertaining to the chest).
> Pertaining to the chest, e.g., pectoral muscles, pectoral appendage.

Pelvic [pel′vik] (L. *pelvis* = basin).
> Pertaining to basin-shaped structures, such as the pelvic girdle, or renal pelvis.

Penis [pē′nis] (L. = tail, penis).
> The male copulatory organ of most amniotes.

Pericardial cavity [per-i-kar′dē-ăl] (Gr. *peri* = around + kardia = heart).
> The portion of the coelom that surrounds the heart.

Pericardium.
> The serosa that covers the surface of the heart (i.e., visceral pericardium) and forms part of the pericardial wall (i.e., parietal pericardium).

Periosteum [per-ē-os′tē-ŭm] (Gr. *osteon* = bone).
> The vascularized and innervated connective tissue covering a living bone.

Peritoneal cavity [per′i-tō-nē-al] (Gr. *peritonaion* = to stretch over).
> The portion of the mammalian coelom that houses the abdominal viscera.

Peritoneum.
> The serosa that covers the abdominal visceral organs (i. e., visceral peritoneum) and lines the peritoneal cavity (i.e., parietal peritoneum).

Peroneus muscle [per-ō-nē′ŭs] (Gr. *perone* = pin, fibula).
> One or more muscles located on the lateral side of the shin over the fibula and extending into the foot.

Phalanges [fă-lan′jēz] (Gr. = battle lines of soldiers).
> Bones of the digits that extend beyond the palm of the hand or sole of the foot.

Pharynx [far′ingks] (Gr. = throat).
> That part of the digestive tract that lies between the oral cavity and the esophagus; the crossing place of the digestive and respiratory tracts. Embryonically, lungs develop as outgrowths from the pharyngeal floor, and pharyngeal pouches develop from the lateral walls of the pharynx.

Phrenic [fren′ik] (Gr. *phren* = diaphragm).
> Pertaining to the diaphragm, e.g., phrenic nerve, phrenic artery.

Pia mater [pī′ă mā′ter] (L. = tender mother).
> The delicate, vascularized layer of connective tissue that invests the brain and spinal cord. It is the innermost of the three mammalian meninges.

Pineal gland [pin′ē-ăl] (L. *pineus* = pertaining to pine, from *pinus* = pine tree).
> An endocrine gland that produces melatonin, especially under dark conditions. The functions of melatonin are not fully understood, but it has been implicated in adjusting physiological processes to diurnal and seasonal cycles.

Piriformis muscle [pir′i-fōrm-is] (L. *pirum* = pear + *forma* = shape).
> A small, triangular muscle located on the medial side of the proximal end of the thigh. It is pear-shaped in human beings.

Pituitary gland.
> See **Hypophysis.**

Placenta [plă-sen′tă] (L. = flat cake).
> The apposition or union of parts of the uterine lining and fetal extraembryonic membranes through which food, respiratory gases, and waste products are exchanged between the mother and the fetus.

Plantaris muscle [plan′tār-is] (L. = pertaining to the sole of the foot).

A large muscle on the caudal surface of the crus of nonmammalian vertebrates. Its tendon runs onto the sole of the foot and extends the foot.

Platelet [plāt′let] (English, *small plate*).

An irregularly shaped fragment of the cytoplasm of a megakaryocyte in the blood. It is involved in blood clotting.

Platysma muscle [plă-tiz′mă] (Gr. *platys* = flat, broad + *-ma* = suffix indicating result of an action, e.g., of making something flat and broad).

Thin sheet of muscle underlying the skin of the neck in mammals.

Pleura [plūr′ă] (Gr. = side, rib).

The serosa that covers the lungs (i.e., visceral pleura) and lines the pleural cavities (i.e., parietal pleura).

Pleural cavities.

The paired coelomic spaces that enclose the lungs in mammals.

Pleuroperitoneal cavity [plūr-ō-per-i-tō-nē′al].

The combined potential peritoneal and pleural cavities in anamniotes. It contains the abdominal viscera and, if present, the lungs.

Plexus [plek′sŭs] (L. = network, braid).

A network of blood vessels or nerves, e.g., choroid plexus, brachial plexus.

Polar body [pō′lăr] (L. *polaris* = pole).

A small body located near the animal pole of a developing egg cell. It results from the unequal division of the egg during the first and second meiotic division.

Pollex [pol′eks] (Gr. = thumb).

The thumb.

Pons [ponz] (L. = bridge).

The ventral part of the mammalian metencephalon. Its most conspicuous surface feature is a bridgelike tract of transverse neuron fibers.

Popliteal [pop-lit′ē-ăl] (L. *poples*, gen. *poplitis* = knee joint).

The depression behind the mammalian knee joint.

Portal vein [pōr′tăl] (L. *porta* = gate, door).

A vein that carries blood from the capillaries of one organ to the capillaries of another organ rather than to the heart, e.g., the hepatic portal vein.

Posterior [pos-tēr′ē-ōr] (L. *poster* = after, following + *ior* = suffix indicating the comparative form of an adjective, e.g., more behind).

A direction toward the back of a human being. It is sometimes used for the tail end of a quadruped, but *caudal* is a more appropriate term.

Posterior chamber.

The space within the eyeball between the iris and lens. It is filled with aqueous humor.

Postganglionic motor neuron [post′gang-glē-on′ik] (Gr. *ganglion* = small tumor, swelling).

A neuron of the autonomic nervous system whose cell body lies in a ganglion in the peripheral nervous system.

Preganglionic motor neuron [prē′gang-glē-on′ik] (L. *pre-* = in front of).

A neuron of the autonomic nervous system whose cell body lies within the central nervous system and whose axon synapses with a postganglionic neuron.

Premolar tooth [prē-mō′lăr] (L. *molaris* = millstone).

One of several mammalian teeth that lie in front of the molar teeth and behind the canine tooth. They usually are adapted for a combination of cutting and grinding.

Prepuce [prē′pūs] (L. *praeputium* = foreskin).

The foreskin of a male mammal. It covers the glans penis.

Prosencephalon [prōs-en-sef′ă-lon] (Gr. *pro-* = before + *enkephalos* = brain).

The embryonic forebrain. It gives rise to the telencephalon and diencephalon.

Prostate [pros′tāt] (Gr. *prostates* = one who stands before).

An accessory genital gland of male mammals. It surrounds the urethra just before the urinary bladder and secretes much of the seminal fluid.

Protraction [prō-trak′shŭn] (L. *traho*, p. p. *tractus* = to pull).

Muscle action that moves the entire appendage of a quadruped forward.

Proximal [prok′si-măl] (L. *proximus* = nearest).

The end of a structure nearest its origin, e.g., the end of a limb that is closest to the girdle.

Pterygoideus muscle [ter′i-goyd-e-us] (Gr. *pteryx, pteryg-* = wing + *eidos* = resembling).

A jaw-closing muscle that arises from the pterygoid bone (frog) or pterygoid process on the underside of the skull (mammals) and inserts on the medial side of the lower jaw.

Pubis [pyū′bis] (L. = genital hair).

The cranioventral bone of the pelvic girdle of terrestrial vertebrates.

Pudendal [pyū-den′dăl] (L. *pudendum* = external genitals, from *pudendo* = to feel ashamed).

Pertaining to the region of the external genitals, e.g., pudendal artery.

Pulmonary [pŭl′mō-nār-ē] (L. *pulmo* = lung).

Pertaining to the lungs.

Pulmonary trunk.
> The mammalian arterial trunk that leaves the right ventricle and soon divides into the two pulmonary arteries going to the lungs.

Pulmonary valve.
> A set of three semilunar folds in the base of the pulmonary trunk. It prevents the backflow of blood into the right ventricle.

Pupil [pyū′pǐl] (L. *pupilla* = pupil).
> The central opening in the iris through which light enters the eyeball.

Pylorus [pī-lōr′ŭs] (Gr. *pyloros* = gatekeeper).
> The caudal end of the stomach, which contains a sphincter muscle.

Pyramidal system [pi-ram′i-dal] (Gr. *pyramis* = pyramid).
> The direct neuronal motor pathway in mammals, from pyramid-shaped cell bodies in the cerebrum to motor neurons in the brain and spinal cord.

Quadrate cartilage [kwah′drāt] (L. *quadratus* = square).
> A cartilage at the caudal end of the upper jaw of a frog with which the mandibular cartilage of the lower jaw articulates. It represents the dorsal half of the first visceral arch of ancestral fishes.

Radius [rā′dē-ŭs] (L. = ray, spoke).
> The bone of the forearm that rotates around the ulna in most terrestrial vertebrates. It lies on the lateral side of the forearm when the palm is supine.

Ramus [rā′mŭs] (L. = branch).
> A branch of a nerve or a blood vessel, also part of the mandible.

Rectum [rek′tŭm] (L. *rectus* = straight).
> The caudal end of the large intestine of a mammal.

Renal [rē′năl] (L. *ren* = kidney).
> Pertaining to the kidney, e.g., renal artery.

Renal portal system.
> A system of veins that drains the capillaries of the hind legs and tail of most nonmammalian vertebrates and leads to capillaries around the kidney tubules.

Renal tubule.
> A kidney tubule. It consists of a glomerular capsule and a tubule; the tubular part of a nephron.

Rete testis [rē′tē tes′tis] (L. = net + *testis* = testicle).
> A network of small passages in the mammalian testis between the seminiferous tubules and the epididymis. They form a visible cord in the rat.

Retina [ret′i-nă].
> The innermost layer of the eyeball. It contains pigment, receptive rods and cones, and associated neurons.

Retraction [rē-trak′shŭn] (L. *re* = backward + *tractio* = a pull).
> Muscle action that pulls the entire appendage of a quadruped caudad.

Rhinal [rī′năl] (Gr. *rhis*, gen. *rhinos* = nose).
> Pertaining to the nose.

Rhinal sulcus.
> The furrow that separates the portion of the cerebrum dealing with olfaction from other areas of the brain.

Rhinencephalon [rī′nen-sef′ă-lon] (Gr. *enkephalos* = brain).
> The primary olfactory portion of the cerebrum.

Rhombencephalon [rom-ben-sef′ă-lon] (Gr. *rhombus* = an oblique-angled equilateral parallelogram + *enkephalos* = brain).
> The embryonic hindbrain. It develops into the metencephalon and myelencephalon.

Rodentia [rō-den′shē-ă] (L. *rodo*, pres. p. *rodens* = to gnaw).
> The mammalian order, including the rat and other gnawing animals, that is characterized by two enlarged incisor teeth in the upper and lower jaws.

Rostral [ros′trăl] (L. *rostrum* = beak).
> A direction toward the front of the head. This term is used for structures within the head.

Round ligament.
> A round cord; specifically, a connective tissue cord that crosses the broad ligament in female mammals and connects the ovary with the body wall at the groin. It corresponds to the male gubernaculum.

Round window.
> See **Fenestra cochleae.**

Sacculus [sak′yū-lŭs] (L. small sac).
> The most ventral chamber of the inner ear. It contains an otolith and functions as a sensor of equilibrium.

Sacral vertebrae, sacrum [sa′krŭm] (L. = sacred).
> The vertebrae, or fusion of two or more vertebrae and their ribs, with which the pelvis articulates.

Sagittal [saj′i-tăl] (L. *sagitta* = arrow).
> A plane of the body that passes through the sagittal suture (between the parietal bones of the skull), e.g., a median, longitudinal plane passing from dorsal to ventral.

Salivary gland [sal′i-vār-ē] (L. *saliva* = saliva).
> One of several glands that secrete saliva. Major ones in mammals are the parotid, mandibular, and sublingual glands.

Sarcolemma [sar′kō-lem-ă] (Gr. *sarx*, gen,. *sarkos* = flesh + *lemma* = husk).
> The cell membrane of a muscle cell.

Sartorius muscle [sar-tōr′ē-ŭs] (L. *sartor* = tailor).
> A narrow, diagonal muscle on the medial surface of the thigh. It is especially well developed in people

who sit cross-legged on the floor, such as tailors in preindustrial times, because it is used to rise.

Scapula [skap′yū-lă] (L. = shoulder blade).
The shoulder blade, or the element of the pectoral girdle that extends dorsally on the back.

Sclera [sklēr′a] (Gr. *skleros* = hard).
The opaque (white) portion of the fibrous tunic of the eyeball. Together with the cornea, it forms the outer wall of the eyeball.

Scrotum [skrō′tŭm] (L. = pouch).
The cutaneous sac that encases the paired mammalian testes when the testes are descended.

Sebaceous gland [sē-bā′shŭs] (L. *sebum* = tallow).
A gland in mammalian skin. It produces an oily or waxy secretion, which is usually discharged into a hair follicle.

Secondary palate.
The palate of mammals that separates the food and air passages. It consists of the hard palate, which separates the oral from the nasal cavities, and the fleshy soft palate, which separates the oral pharynx from the nasal pharynx.

Semicircular duct [sem′ē-sir-kyū-lăr] (L. *semi* = prefix denoting one-half).
One of three semicircular ducts in the inner ear. Each lies at a right angle to the others and detects changes in angular acceleration generated by turns of the head.

Seminal fluid [sem′i-nă l] (L. *semen* = seed).
The fluid secreted by male reproductive ducts and accessory genital glands.

Seminal vesicle.
See **Vesicular gland.**

Seminiferous tubules [sem-i-nif′er-ŭs] (L. *semen,* gen. *seminis* = seed + *fero* = to carry).
Tubules within the testis in which sperm cells are produced.

Septum [sep′tum] (L. = partition).
A partition.

Septum pellucidum [pe-lū′si-dum] (L. *pellucidus* = clear, transparent).
A thin, vertical septum of nervous tissue in the brain ventral to the corpus callosum. It forms the medial wall of each lateral ventricle.

Serosa [se-rō′să] (L. *serosus* = watery).
The coelomic epithelium and underlying connective tissue that line body cavities and cover visceral organs, e.g., peritoneum, pleura, pericardium, and tunica vaginalis.

Serratus muscle [ser-āt′us] (L. = toothed like a saw).
A muscle with a serrated or saw-toothed border.

Sertoli cell (*Enrico Sertoli,* Italian histologist, 1842–1910).
Large epithelial cell in the seminiferous tubules. It plays a role in the maturation of the sperm cells.

Sesamoid bone [ses′ă-moyd] (Gr. *sesamon* = sesame seed + *eidos* = resembling).
A bone that develops in the tendon of some muscles near their insertion and facilitates the movement of the tendon across a joint, or changes the direction of the force of a muscle, e.g., the patella and the pisiform.

Sinus [sī -nŭs] (L. = a cavity).
A cavity or space within an organ.

Sinus venosus [vē-nō′sus] (L. *venosus* = pertaining to a vein).
The most caudal chamber of the heart of anamniotes and some reptiles. In frogs, it receives blood from the body and leads to the right atrium.

Skin.
See **Integument.**

Skull [skŭl] (Old English *skulle* = a bowl).
The group of bones and cartilages that encase the brain and major sense organs and form the upper jaw and face. The lower jaw sometimes is considered to be a part separate from the skull.

Soft palate.
The fleshy partition in mammals between the oral pharynx and nasal pharynx.

Somatic [sō-mat′ik] (Gr. *somatikos* = bodily).
Pertaining to the body wall and appendages, as opposed to the internal (visceral) organs, e.g., somatic muscles, somatic skeleton.

Sperm cell [sperm] (Gr. *sperma* = seed, sperm cell).
One of the mature male gametes; also called spermatozoa.

Spermatid [sper′mă-tid] (Gr. *idion,* a diminutive ending).
A cell resulting from the second meiotic division of a secondary spermatocyte. It develops into a sperm cell.

Spermatocyte [sper′mă-tō-sīt] (Gr. *kytos* = cell).
The primary spermatocyte results from the growth and division of a spermatogonium, and the secondary spermatocyte results from the first meiotic division of the primary spermatocyte.

Spermatogonium [sper′mă-tō-gō-nē-ŭm] (Gr. *gone* = generation).
The stem cell that multiplies mitotically and gives rise to sperm forming cells.

Spermatozoa.
See **Sperm cell.**

Spinal [spī ′năl] (L. *spina* = spine, thorn).
Pertaining to a spine-shaped structure, often the spine or vertebral column, e.g., spinal cord, spinal nerve.

Spinous process.
The dorsal, spinelike process of the vertebral arch.

Splanchnic [splangk′nik] (Gr. *splanchnon* = gut, viscus).
Pertaining to structures that supply the gut or visceral organs, such as the splanchnic nerve.

Spleen [splēn] (Gr. *splen* = spleen).

A large lymphoid organ located near the left side of the stomach. In various vertebrates and at different times in their life cycle, it produces, stores, or disassembles senescent red blood cells.

Splenius muscle [splē′nē-ŭs] (Gr. *splenion* = bandage).

A thin, triangular sheet of muscle located on the back of the neck of mammals deep to part of the rhomboideus muscle.

Stapes [stā′pēz] (L. = stirrup).

The single auditory ossicle of nonmammalian vertebrates, and the innermost of the three auditory ossicles in mammals.

Statoacoustic nerve [stat′ō-ă-kū′stik] (Gr. *statos* = standing still + *akoustikos* = pertaining to hearing).

A name often used for the eighth cranial nerve in nonmammalian vertebrates. It is called the vestibulocochlear nerve in mammals.

Sternum [ster′nŭm] (Gr. *sternon* = chest).

The breastbone of terrestrial vertebrates.

Stomach [stŭm′ŭk] (Gr. *stomakos* = stomach).

Saclike part of the digestive tract lying between the esophagus and intestine in which food is temporarily stored and digestion usually is initiated.

Stratum [strat′ŭm] (L. = layer).

A layer of tissue, such as the stratum corneum of the skin.

Styloid process [stī′loyd] (Gr. *stylos* = pillar + *eidos* = resembling).

A slender process on the ventral surface of the skull of many mammals. The hyoid apparatus attaches to it.

Subclavian [sŭb-klā′vē-an] (L. *sub-* = beneath + *clavis* = key).

Pertains to a position beneath the clavicle, e.g., the subclavian artery.

Sublingual gland [sŭb-ling′gwăl] (L. *lingua* = tongue).

A salivary gland of mammals. It lies beneath the tongue.

Submucosa [sŭb-mŭ-kō′să] (L. *mucosus* = mucous).

A layer of vascularized connective tissue in the wall of the digestive or respiratory tract. It underlies the mucosa.

Sulcus [sŭl′kŭs] (L. = groove).

A groove on the surface of an organ, e.g., the sulci on the surface of the cerebrum of many mammals.

Superior [sū-pēr′ē-ōr] (L. *super* = above + *-ior* = suffix indicating the comparative form of an adjective, e.g., more above).

A direction toward the head of a human being.

Suprarenal gland [sū-pră-rē′năl].

See **Adrenal gland.**

Suture [sū′chūr] (L. = seam).

An essentially immovable joint in which the bones are separated by connective tissue, e.g., many of the joints between bones of the skull.

Sympathetic nervous system [sim-pă-thet′ik] (Gr. *syn* = with + *pathos* = feeling).

The part of the autonomic nervous system that, in mammals, leaves the central nervous system from the thoracic and lumbar parts of the spinal cord. Its activity helps an animal adjust to stress by promoting physiological processes that mobilize the energy available to the tissues.

Symphysis [sim′fi-sis] (Gr. *physis* = a growth).

A joint or fusion between bones that permits limited movement by the deformation of fibrocartilage between them. It occurs in the midline of the body, e.g., the pelvic symphysis, mandibular symphysis.

Systemic [sis-tem′ik] (Gr. *systema* = whole).

Pertaining to the body as a whole rather than to a specific part, e.g., the systemic circulation as opposed to the pulmonary circulation.

Talus [tā′lus] (L. = ankle bone).

The proximal tarsal bone of mammals. It articulates with the tibia.

Tapetum lucidum [tă-pē′tŭm lū′sid-ŭm] (L. = carpet + *lucidus* = shining).

A layer within or behind the retina of some vertebrates. It reflects light back onto the photoreceptive cells.

Tarsal [tar′săl] (Gr. *tarsos* = ankle).

One of the small bones of the ankle.

Tela choroidea [tē′la kōr-oy′dē-a] (L. *tela* = web, Gr. *chorion* = membrane encasing the fetus + *eidos* = resembling).

A thin membrane forming the roof or part of the wall of some ventricles of the brain. It is composed of the ependymal epithelium and the pia mater.

Telencephalon [tel-en-sef′ă-lon] (Gr. *telos* = end + *enkephalos* = brain).

The most rostral of the five brain regions. It includes the olfactory lobes and cerebrum.

Temporal [tem′pŏ-răl] (L. *tempus*, gen. *temporis* = time).

Pertaining to the temporal region of the skull, so called because the hair in this region is the first to become gray in human beings.

Temporal fossa.

A depression on the lateral surface of the mammalian skull caudal to the orbit and dorsal to the zygomatic arch. It lodges the temporal jaw muscle.

Tendon [ten′dŏn] (L. *tendo* = to stretch).

A band of dense connective tissue attaching a muscle to a bone, or sometimes to another muscle.

Tendon of Achilles (*Achilles*, the hero of Homer's *Iliad* was said to be invulnerable except for his heel).

The tendon that extends from the large muscle mass on the caudal surface of the crus to the calcaneus. These muscles are powerful extensors of the foot.

Tentorium [ten-tō′rē-um] (L. = tent, from *tendo*, p. p. *tentum* = to stretch).

A septum of dura mater in mammals located between the cerebrum and cerebellum. It ossifies in some species.

Teres [ter′ēz] (L. = rounded, smooth).

Descriptive of a round structure, such as the teres muscles.

Tertiary follicle.

A mature follicle in the ovary; also called a *Graafian follicle*.

Testis [tes′tis] (L. = witness, in ancient Rome necessarily an adult male).

The male gonad. It produces sperm cells and the hormone testosterone.

Tetrapod [tet′ră-pod] (Gr. *tetra* = four + *pous*, gen. *podos* = foot).

A collective term for terrestrial vertebrates. They have four feet unless some have been secondarily lost or modified.

Thalamus [thal′ă-mus] (Gr. *thalamos* = inner chamber, bedroom).

The lateral walls of the diencephalon. It is an important center between the cerebrum and other parts of the brain.

Theca [thē′kă] (Gr. *theke* = box, case).

A case or covering, such as the group of connective tissue cells that form the outer layer of an ovarian follicle.

Thoracolumbar fascia [thōr′ă-kō-lŭm′bar].

A part of the deep fascia on the dorsal surface of the thoracic and lumbar regions. It is associated with the muscles of the back and abdominal wall.

Thorax [thō′raks] (Gr. = chest).

The region of the mammalian body encased by the ribs and sternum.

Thymus [thī′mŭs] (Gr. *thymos* = thyme, thymus; so called because of its resemblance to a bunch of thyme).

A lymphoid organ in the ventral part of the neck and thorax. It is essential for the maturation of T-lymphocytes and probably other parts of the immune system. The gland is best developed in young individuals and atrophies later in life.

Thyroid gland [thī′royd] (Gr. *thyreos* = oblong shield + *eidos* = resembling).

An endocrine gland that is usually located near the cranial end of the trachea, but lies over the thyroid cartilage of the larynx in human beings (whence its name). Its hormones increase the rate of metabolism.

Tibia [tib′ē-ă] (L. = the large shinbone).

The large bone on the medial side of the lower leg.

Tissue [tish′ū] (Old French *tissu* = cloth).

An aggregation of cells that together perform a common function.

Tongue [tŭng] (Old English *tunge* = tongue).

A muscular organ in the floor of the oral cavity that often helps gather food and manipulates it within the mouth cavity.

Tonsil [ton′sil] (L. *tonsilla* = tonsil).

One of the lymphoid organs that develop in the wall of the mammalian pharynx.

Trachea [trā′kē-ă] (Gr. *tracheia* = rough artery).

The respiratory tube between the larynx and the bronchi.

Tract [trakt] (L. *traho*, p.p. *tractus* = to pull).

(1) A linear group of organs having a similar function, e.g., the digestive tract. (2) A group of axons of similar function traveling together in the central nervous system.

Transverse [trans-vers′] (L. *transversus* = transverse).

A plane of the body crossing its longitudinal axis at right angles.

Transverse process.

A process of a vertebra, that lies in the transverse plane. Either the tubercle of a rib articulates with it, or an embryonic rib becomes incorporated in it and serves as an attachment site for muscles.

Trapezoid body [trap′ě-zōyd] (Gr. *trapezoides* = resembling a trapezium).

An acoustic commissure at the rostral end of the ventral surface of the mammalian medulla oblongata.

Trigeminal nerve [trī-jem′i-năl] (L. *trigeminus* = threefold).

The fifth cranial nerve. It has three branches in mammals. It innervates the jaw muscles and returns sensory fibers from the surface of the head and oral cavity, except for taste buds.

Trochanter [trō-kan′ter] (Gr. = a runner).

One of the processes on the proximal end of the femur to which certain pelvic and thigh muscles attach.

Trochlear nerve [trok′lē-ăr] (L. *trochlea* = pulley).

The fourth cranial nerve. It innervates one of the extrinsic muscles of the eye, which, in mammals, passes through a connective tissue pulley before inserting on the eyeball.

Truncus arteriosus [trŭng′kŭs ar-ter′ē-ō-sus] (L. = trunk, stem).

One of two arterial trunks in the frog leading from the cranial end of the heart to arterial arches supplying the skin and lungs, head, and body.

Glossary
TEN–TRU

133

Tunic [tū′nik] (L. *tunica* = a coat, covering).
> Descriptive of a layer of an organ, such as one of the layers of the eyeball.

Tunica albuginea [tū′ni-kă al-byū-jin′ē-ă] (L. *albugineus*, from *albugo* = white spot).
> A white, fibrous capsule, such as the one forming the wall of the testis and sending septa into the testis.

Tympanic [tim-pan′ik] (L. *tympanum* = drum).
> Pertaining to the middle ear.

Tympanic cavity.
> The cavity of the middle ear that lies between the tympanic membrane and the inner ear within the otic capsule. One or three auditory ossicles traverse it, and the auditory tube connects it with the pharynx.

Tympanic membrane.
> The eardrum.

Ulna [ŭl′nă] (L. = elbow bone).
> One of the bones of the antebrachium of terrestrial vertebrates. It extends behind the elbow in mammals and lies on the medial side of the antebrachium when the hand is supine.

Umbilical [ŭm-bil′i-kăl] (L. *umbilicus* = navel).
> Pertaining to the navel, e.g., umbilical cord, umbilical artery.

Ureter [yū-rē′ter] (Gr. *oureter* = ureter, from *ouron* = urine).
> The duct in amniotes that carries urine from the kidney to the urinary bladder.

Urethra [yū-rē′thră] (Gr. *ourethra* = urethra).
> The duct that carries urine from the urinary bladder to the cloaca or to the outside of the body in amniotes; part of it also carries seminal fluid and sperm cells in males.

Urinary bladder [yūr′i-nār-ē] (Gr. *ouron* = urine).
> A saccular organ in terrestrial vertebrates in which urine from the kidney accumulates before being discharged from the body.

Urogenital [yū′rō′jen-i-tăl] (L. *genitalis* = creative, fruitful).
> A combining term for the urinary and genital systems, which share many ducts.

Urostyle [yū′rō′stĭl] (Gr. *oura* = tail + *stylos* = a pillar, peg).
> The rodlike, caudal part of the vertebral column of a frog. It is composed of several fused caudal vertebrae.

Uterine tube [yū′ter-in] (L. *uterus* = womb).
> One of a pair of narrow tubes in mammals. It extends from the vicinity of the ovary to the uterus and carries fertilized eggs to the uterus. Also called the *Fallopian tube.*

134

Uterus [yū′ter-ŭs].
> The organ in females in which embryos develop in live-bearing species. It develops from part of the oviduct.

Utriculus [yū′trik′yū-lŭs] (L. = small sac).
> The upper chamber of the inner ear to which the semicircular ducts attach.

Vagina [vă-jī′na] (L. = sheath).
> The part of the mammalian female reproductive tract that receives the penis during copulation.

Vaginal cavity.
> The part of the coelom that contains the testis of male mammals.

Vaginal vestibule [ves′ti-būl] (L. *vestibulum* = antechamber).
> The passage or space into which the vagina and urethra enter in female mammals. It is very shallow in human beings but often forms a short canal in quadrupeds.

Vagus nerve [vā′gŭs] (L. = wandering).
> The tenth cranial nerve. It carries motor fibers to muscles of the larynx and parasympathetic fibers to thoracic and abdominal organs; it returns sensory fibers from these parts of the body.

Vasa efferentia.
> See **Efferent ductules.**

Vascular tunic.
> The middle layer of the eyeball. It forms the choroid, ciliary body, and iris.

Vas deferens.
> See **Ductus deferens.**

Vastus [vas′tus] (L. = great, large).
> Descriptive of some thigh muscles, e.g., the vastus lateralis.

Vein [vān] (L. *vena* = vein).
> A blood vessel that carries blood toward the heart. Usually the blood is low in oxygen content, but pulmonary veins from the lungs have blood with a high oxygen content.

Vena cava [vē′nă cā′vă] (L. = hollow vein).
> One of the primary veins of frogs and amniotes. It leads directly to the heart.

Ventral [ven′tral] (L. *ventralis* = ventral, from *venter* = belly).
> A direction toward the underside of a quadruped.

Ventricle [ven′tri-kl] (L. *ventriculus* = small belly).
> (1) A chamber of the heart. It greatly increases blood pressure and sends the blood to the lungs or to the body. (2) One of the cavities within the brain.

Vermis [ver′mis] (L. = worm).
> The "segmented" and wormlike median portion of the mammalian cerebellum.

Vertebra [ver'te-bră] (L. = vertebra, joint).
One of the units that make up the vertebral column.

Vertebral arch.
The arch of a vertebra that surrounds the spinal cord. It is also called a *neural arch.*

Vertebral body.
The main, supporting component of a vertebra. It lies ventral to the vertebral arch. Also called a vertebral centrum.

Vesicular gland [vĕ-sik'yū-lăr] (L. *vesicula* = small bladder).
One of the accessory genital glands of male mammals that contributes to the seminal fluid; also called the *seminal vesicle.*

Vestibulocochlear nerve [ves-tib'yū-lō-kok-lē-ăr] (L. *cochlea* = snail shell).
The eighth cranial nerve. It returns sensory fibers from the parts of the inner ear related to equilibrium (vestibular apparatus) and from the part monitoring sound detection (cochlea). Often called the statoacoustic nerve in anamniotes, in which a cochlea is absent.

Vibrissae [vī-bris'ē] (L. = vibrissae, from *vibro* = to quiver).
Long, tactile hairs on the snout of many mammals.

Villi [vil'i] (L. = shaggy hair).
Multicellular, but minute, often finger-shaped projections of an organ. They increase its surface area, e.g., the intestinal villi.

Visceral [vis'er-ăl] (L. *viscus*, pl. *viscera* = internal organs).
Pertaining to the inner part of the body as opposed to the body wall and appendages, e.g., visceral muscles, visceral skeleton.

Visceral arches.
The skeletal arches that develop in the wall of the pharynx and may contribute to the formation of the jaws (frog) and parts of the skull, hyoid, and larynx.

Vitreous body [vit'rē-ŭs] (L. *vitreus* = glassy).
The clear, viscous material in the eyeball between the lens and retina.

Viviparous [vī-vip'ă-rŭs] (L. *vivus* = living + *pario* = bearing).
Live-bearing. A pattern of reproduction found in most mammals and a few other vertebrates in which the young develop in a uterus and are born fully formed.

Vocal cords [vō'kăl] (L. *vocalis* = pertaining to voice).
Folds of mucous membrane within the larynx of frogs and mammals. Involved in the production of sounds.

Vomer bone [vō'mer] (L. = plowshare).
A bone in the roof of the mouth in frogs and in the floor of the nasal cavities in mammals. It is plowshare-shaped.

Vomeronasal organ [vō'mer-ō-nā-zăl].
An accessory olfactory organ located between the palate and the nasal cavities of most terrestrial vertebrates. It is important in feeding and sexual behaviors. It is also called **Jacobson's organ.**

Vulva [vŭl'vă] (L. = covering).
The female external genitalia.

Wharton's jelly (*Thomas Wharton,* British anatomist and physician, 1614 – 1673).
The mucoid connective tissue of the umbilical cord.

White matter.
Tissue in the central nervous system that consists primarily of myelinated axons.

Wolffian duct (*Kaspar Friedrich Wolff,* German embryologist in Russia, 1733–1794).
See **Archinephric duct.**

Yolk sac [yōk] (Old English *geulca* = yolk, from *geolu* = yellow).
The yolk-containing sac attached to the ventral surface of the embryo in some fishes, reptiles, birds, and egg-laying mammals. It is reduced in viviparous mammals.

Zonule fibers [zō'nyūl] (L. *zonula* = small zone).
Delicate fibers extending between the ciliary body and the lens equator. These fibers transmit forces from the ciliary body to the lens.

Zygomatic arch [zī'gō-mat-ik] (Gr. *zygoma* = bar, yoke).
The arch of bone beneath the orbit in a mammalian skull. It connects the facial and cranial regions.